D0209086

Concrete Jungle

*The publisher gratefully acknowledges the
generous support of the Humanities Endowment
Fund of the University of California Press Foundation.*

*The publisher also gratefully acknowledges the generous support
of the Ralph and Shirley Shapiro Endowment Fund in Environmental
Studies of the University of California Press Foundation.*

Concrete Jungle

*New York City and Our Last Best Hope
for a Sustainable Future*

Niles Eldredge and Sidney Horenstein

UNIVERSITY OF CALIFORNIA PRESS

University of California Press, one of the most
distinguished university presses in the United States,
enriches lives around the world by advancing scholarship
in the humanities, social sciences, and natural sciences. Its
activities are supported by the UC Press Foundation and
by philanthropic contributions from individuals and
institutions. For more information, visit www.ucpress.edu.

University of California Press
Oakland, California

Library of Congress Cataloging-in-Publication Data

Eldredge, Niles, author.
    Concrete jungle : New York City and our last best
hope for a sustainable future / Niles Eldredge and
Sidney Horenstein.
        pages   cm
    Includes bibliographical references and index.
    ISBN 978-0-520-27015-2 (cloth : alk. paper)
    ISBN 978-0-520-95830-2 (ebook)
    1. Urban ecology (Sociology)—New York (State)—
New York.   2. Urban geography—New York (State)—
New York.   3. Environmental degradation—New
York (State)—New York.   4. Biodiversity—New York
(State)—New York.   I. Horenstein, Sidney S.,
author.   II. Title.
    HT243.U62N736   2014
    307.7609747—dc23

                                    2014014646

Manufactured in the United States of America

23 22 21 20 19 18 17 16 15 14
10 9 8 7 6 5 4 3 2 1

The paper used in this publication meets the minimum
requirements of ANSI/NISO z39.48–1992 (R 2002) (*Permanence
of Paper*).

CONTENTS

The Yin and Yang of Cities

The story of cities and their relation to the natural world is a mass of contradictions. Yet our basic thesis is relatively simple: cities are the most extreme examples of environmental destruction, because the very act of their construction utterly destroys habitats, and because their voracious appetites for resources such as food and water extend far beyond their limits and, these days, all over the world. Yet ironically and simultaneously, cities represent the last best hope for conserving healthy remnants of the world's species and ecosystems—systems vital to the continued existence of viable life, certainly including human life, on earth.

Cities have done a lot to clean up their environmental act in recent decades, and the general green movement that many cities have begun to follow has improved their internal environments considerably: the air is generally fresher, at least in some places; the water safer; and parks and gardens have begun to be restaffed by species (some native, some alien) that had all but disappeared from urban settings. And the recent work of urban planners, architects, and others concerned with the ultimate fates of cities has made a convincing case that what might be best for the still-burgeoning human population would be to concentrate proportionately even greater numbers of us into cities, thereby providing at least a modicum of relief to the seemingly unending development of rural and wild areas that inevitably results in their degradation.

But there is more to the plus side of cities in relation to the natural world. By their very nature, by their very concentration of human beings, what cities bring to the table as a source of at least partial environmental global salvation is their cultural resources in the broadest sense: their centers of learning, research, government, finance, media—all sorts of organizations and institutions that increasingly play active local, national, and international roles in solving environmental problems. Cities now are reaching out and giving back as much as they are taking away and, in some cases, ameliorating environmental destruction.

Both of us are New Yorkers born and bred. We love our city and have learned much about it, particularly through the fact that we belong to a relatively rare breed: we are native New Yorkers who have spent our professional lives studying what used to be called natural history—the geological and biological history of the earth, including the physical environment, the ecosystems, and the species around us today. Indeed, we have worked side by side at the American Museum of Natural History since the 1960s (though we are both now retired). Together with other colleagues, we worked to develop the first "issues" hall at the American Museum: the Hall of Biodiversity, which opened in the mid-1990s. And though much of our work has involved research in paleontology (the history of ancient and long-since extinct species), each of us has branched out, extending our interests in rocks and fossils to a variety of environmental issues in the modern world.

One of us (SH) has become a well-known figure in local New York environmental circles, concentrating many of his efforts on issues such as that of water resources and quality and on its counterpart, the cleaning of polluted waterways and the general problem of sanitation. He has been a geology and environmental consultant for many organizations (see this volume's acknowledgments). He knows all the parks and green spaces and their history intimately. And he has walked all the streets of Manhattan and many of the surrounding boroughs and has led trips all over the New York cityscape. His photos and personal vignettes included in our text bring to life the environmental and geologic story of New York and help us, at the same time, to see that New York, unique though it is in many ways, is nonetheless representative of cities the world over in terms of its path of environmental destruction, then amelioration. Also unique is its development of significant institutions that we feel lie at the heart of ongoing efforts to stave off further global environmental damage.

The other of us (NE) has been interested in evolutionary biology throughout his career, developing (with the late Stephen Jay Gould) the theory of "punctuated equilibria" in the late 1960s and early 1970s. By the 1980s, it had become undeniably clear (largely through the efforts of Norman D. Newell, mentor to both Gould and Eldredge and chairman of the American Museum department in which Horenstein and, a bit later, Eldredge spent their working careers) that nothing much happens in evolution without prior environmental shake-ups and extinctions of preexisting species, sometimes including monumental global mass extinction events. The obvious direction was to ask what past extinction events could tell us about modern-day extinctions: the "Sixth Extinction" that had become all too clear was (and still is) engulfing the species of the modern world. That of course meant learning about the modern extinction crisis, which led to publication of three earlier books and to the work with Horenstein and others on the museum's Hall of Biodiversity.

In a nutshell, we take environmental problems extremely seriously and have made it our business to learn as much about the yin and yang of the thorny relationship between cities and the natural world. Our special knowledge of, and love for, New York has provided us with a template for exploring these issues in this book. Read it for the message, by all means. But read it, too, for the sketch of New York's environmental history we provide.

Finally, for ease of reading, we have cited our sources, including sources for quotations, in a single section—Notes, References, and Suggestions for Further Reading—at the end of our narrative.

CHAPTER ONE

Regarding Broadway

The Urban Saga and the New York Microcosm

Times Square pulses twenty-four hours a day. Flashing neon signs light up the night sky as crowds of New Yorkers and tourists dodge cars and buses, despite occasional crackdowns on jaywalking. We may find similar displays in the streets of Shanghai, London, Paris, and Tokyo, but somehow it is New York, with its Forty-Second-Street-and-Broadway anchor point for Times Square, that seems more than any other to be the crossroads of the world.

The new millennium brought a sanitization effort to the tawdrier side of Times Square, with sex shops, streetwalkers, and X-rated movies shunted downtown or to peripheral streets and avenues. Who knows how long this "Disneyfication"—with its spate of glass-and-steel hotels, and its theaters proffering "wholesome family entertainment"—will last before it too is replaced by the next wave of change, whether this means still-newer forms of entertainment or the lapse back into sleaze that would reflect yet another economic downturn.

Whatever happens, one thing is sure: nothing stays the same. Cities are in constant flux. When George M. Cohan wrote "Give My Regards to Broadway" in 1904, he followed the first line with "Remember me to Herald Square. Tell all the folks on 42nd Street that I will soon be there"—a reminder of the days when the theater district was in the act

Figure 1. Times Square, viewed from Duffy Square to the south. Named for the headquarters of the *New York Times*, Times Square is also called the crossroads of the world and the Great White Way. In 1913 the junction of Forty-Second Street and Broadway, at the southern border of the square, became the eastern terminus of the Lincoln Highway, the first road across the United States. Photograph by Sidney Horenstein.

of pulling up its roots at Thirty-Fourth Street (that is, Herald Square, home to Macy's) and about to reinvent itself in the Times Square neighborhood, since known simply as *the* theater district.

Beyond recognizing the flux of a city's normal growth and change, we can also see that cities, like people, have births, lifetimes, and inevitably declines and demises. What is more, there simply was no such thing as a city just ten thousand years ago. Back then, there were roughly 5 million people on the planet—maybe 10 million at most. By 2011, according to U.S. Census Bureau estimates, New York City alone had some 8,175,133 residents, and millions more if you count all of Greater New York. What's more, those 5 or 10 million human souls on earth ten thousand years ago were thinly distributed around the globe, all the way from present-day Cape Town in southernmost Africa, up through the environs of modern Paris and Berlin, east over to where Calcutta and Beijing now stand, then on down to the future sites of Bangkok in Thailand and Jakarta in Indonesia, and all the way to where Australia's Adelaide, Perth, and Sydney were to be built many millennia later. People had also reached the Western Hemisphere, but just barely. By then, they might already have traversed what would someday become Manhattan Island. And while some bands had reached as far south as modern-day Brasilia, people hadn't yet made it to Kingston in the Caribbean islands.

Only small, temporary villages held any concentration of people, for humans, scattered around the globe as they were, still lived the semi-nomadic life that is the lot of all folk who hunt, fish, and gather wild plants to stay alive. The relatively few hunter-gatherers who survive today generally cluster in small groups of between 40 and 70 people. These bands live in temporary encampments with hide tents or simple thatch or wooden huts, migrating seasonally or as environmental conditions dictate. It cannot be otherwise, for all their water and food supplies come from their surrounds, the natural ecosystems of which they are an integral part.

Think of a hunter-gatherer, a San person (a "Bushman"), say, living in the Kalahari Desert that sprawls across large sections of Botswana,

Figure 2. Former display at the American Museum of Natural History. This diorama depicts a Native American encampment at the northern end of Manhattan Island. Negative #282788, courtesy of the Division of Anthropology, American Museum of Natural History.

Namibia, and South Africa. The distinction between life in a semi-desert and life in millennial New York may seem obvious: Aboriginal San, documented by anthropologists, had few material goods beyond their simple clothing: bows and poison arrows, gourd bowls, and a few other products fashioned from the minerals, plants, and animals that came to hand.[1] New Yorkers of all social strata, however, have televisions, computers, cell phones, electricity, hot and cold running water, and so on, an almost limitless list of material goods inextricably linked to modern life. New Yorkers live in brick-and-mortar, steel-and-concrete, wooden, or granite or brownstone buildings, all of which exude

1. No group of hunter-gatherers, including the San, remains untouched by contact with the Western, technologized world, as anyone who has seen *The Gods Must Be Crazy* will realize. San peoples today live as hunter-gatherers only part of the year, if at all; for the most part, they sport Western clothing and, when living in permanent settlements, have begun to acquire as much of the paraphernalia of modern industrial nations, including cars and TV sets, as their economic resources will allow.

Figure 3. Local market offering fresh fruit and other foods from countries around the world. Beginning in the early 1990s, the amount of imported food sold in local markets increased substantially. Photograph by Sidney Horenstein.

an aura, at least, of permanence; San, in contrast, lived in easily constructed and dismantled one-room huts.

More crucially and tellingly, San men were often out on the hunt, while San women were constantly collecting tubers, melons, and other plants to prepare their meals. All members of the group were highly aware of where their food and water came from. In stark contrast, New Yorkers go shopping for their supplies at the corner market, the big supermarket, or upscale food emporia. New Yorkers in general think no more of where their food comes from (other than "da store") than they think about where their electricity, running water, or heat comes from.

But below the surface differences between the San way of life and a New Yorker's lies a profound gulf of radical change. The New York mode of getting food from the store (much less takeout delivery) did not somehow evolve gradually from the constant search for food that was the human condition until around ten thousand years ago. Instead,

close to home. Throughout history, farming has led to increasingly permanent settlements. Long gone are the days when small villages of farmers could simply pull up stakes and find some new locale in which to set up their enterprise, for as human numbers have grown, available land has proportionately shrunk. And these settlements have of course grown—to the point where no one thinks of a city like New York as a place where farmers live and go out every day to tend to their crops in adjoining fields. Never mind that until around 1900, Brooklyn (the third-largest city in the United States until 1898, when it became a borough of the city of Greater New York) supplied much of the green groceries that reached the dining tables of New York City—meaning, for the most part, the denizens of Manhattan Island. Truck farming was a vibrant feature of the landscape of the Bronx, Staten Island, and Queens, not to mention a feature of what later became, after World War II, the rural suburbs of Westchester and, increasingly, of New Jersey's coastal plain. (New Jersey's nickname, "the Garden State," pays tribute to its former role in provisioning "the Empire State.")

Agriculture first arose in several different world regions at once, apparently independently, as if it were an idea whose time had come. The earliest records are from the Middle East, where the Natufian peoples of the Levant region (bordering the eastern Mediterranean) seem to have been the first to master the rudiments of isolating kernels of primitive wheat (traditionally called "corn" by Europeans, a source of great confusion to speakers of American English) and sowing them for next year's crop. This was about ten thousand to eleven thousand years ago. The famous Fertile Crescent, now largely the site of Iraq—a land bounded by the Tigris and Euphrates Rivers—was another early locus of farming, as was the incredibly fertile Nile Valley, whose fields were replenished every year by the floods that rise up deep in the heart of the African continent. With the floods came nutrients and a thin yet rich layer of new soil—no crop rotation necessary in Egypt!

No one supposes that these several, separate sites in the Middle East all reflect independent invention of farming know-how. Though war-

Figure 4. One of the last farms on Manhattan Island, in a photo taken about 1890 in what is now Inwood Hill Park. Farmer heritage in this area goes back to Dutch times, when the Nagel and Dyckman families owned much of this part of the island. Bolton Collection, courtesy of the Division of Anthropology, American Museum of Natural History.

fare was common among the early states, so was diplomatic contact that fostered the spread of knowledge. Even warfare brought innovation, as conquering peoples brought with them new ideas. Probably the most famous example was the introduction of the wheeled vehicle to Egypt by the mysterious Hyksos invaders. (This in contrast to records of the potter's wheel seen in Egyptian tomb pictures from about 2500 B.C.) The Hyksos, also referred to as Hyk-Sos or as "shepherd kings," defeated the Egyptians and closed the pages of history on the "Middle Kingdom," largely by dint of their use of horse-drawn chariots, which until then had been wholly unknown to Nile Valley dwellers. The Hyksos invasion came at about 1650 B.C., after some two thousand years of successful Egyptian life, when cities like Memphis flourished and the huge pyramids were built, all without the aid of wheels.

There was plenty of communication to spread the word about farming around the Middle East (and eventually around the entire Mediterranean basin, and later into western Europe). But people elsewhere, seemingly independently, were also discovering the advantages of taking control by planting crops and domesticating animals for food and other uses, as hunting aids and home protection, not to mention as friendly household pets. Along the Indus River in what is now Pakistan, the city of Mohenjo-daro sprang up, along with the first flickers of what was to develop into the vast, rich Indian culture. Agriculture also fueled the early fires of Chinese civilization, at first especially along the banks of what Anglophones now call the Yellow and Yangtze Rivers.

The people of the New World, who had first arrived from Asia in significant numbers starting around 12,500 years ago (though some evidence suggests earlier arrival dates for the very first immigrants), also developed several farming nuclei. Most famous are the Mayan and Aztec city-states of Central America and Mexico, as well as the cities of the Incan peoples of Peru. But agriculture soon developed virtually everywhere in the Americas, as the pilgrims and other early settlers of the east coast of North America quickly discovered.

One thing is clear: although we tend to think of farming as a purely rural activity, far removed from the inner workings of major cities, it is certain that without agriculture there would be, indeed there *could* be, no cities. True, many agricultural peoples have lacked a major settlement you would call a city. And many rural areas, even in technologically advanced nations like the United States, are far away from anything remotely resembling a true city. But the reverse is not true: there never was a city unless there was agriculture.

Looking at any map reveals another major feature of cities: by far the great majority of them are close to water, situated either along riverbanks, on the shores of major lakes, or—among those founded somewhat later—on harbors opening to the sea. Today, eight of the ten largest cities in the world are adjacent to, or connected to, the sea (the exceptions are Delhi and Mexico City). Most fertile farming areas, after all, are in the floodplains of rivers, so it was only natural that the earliest agriculturalists would be living where they had the best chance to succeed at this grand experiment in cultivation. And of course, farmers need water for their fields. The annual flood of the Nile brought more than nutrients and a thin rime of fertile new topsoil: it brought water itself, which the early Egyptian farmers learned to trap in a system of canals and dams. During the dry season, the Egyptian farmers would lift the water up to the level of the fields with a number of clever devices—including a simple clay pot on a counterbalanced pole (the shadouf) and, somewhat later, the Archimedes screw, a large screw inserted inside a cylinder that drew water upward when the screw was rotated. Still later, the Egyptians were bequeathed the ox-driven water wheel by their Roman conquerors.

Water is so important to the agricultural enterprise that some scholars have insisted that control of the waterways was the key to political control ("unification") of large areas, typically along the course of major rivers such as, once again, the Nile. There's a lot to this argument: consider the political troubles over water rights that continue to beset the American West. Much of the water of eastern California and western Nevada was long ago earmarked for thirsty Los Angeles; with urban growth outstripping even these supplies, Los Angeles has turned to the waters of the Colorado River—which now no longer reaches the Gulf of California.

Nor are New Yorkers immune from political hassles over water, as we shall see in greater detail in chapter 5. New York's water supply lies well to the north, and while New York City might own the land around its upstate reservoirs and along the rights-of-way of its pipelines in order to protect the drainage basin and control sources of pollution, there has been friction over water use over the years, mitigated in recent years by cooperative policies. Restrictions on reservoir usage is but one of the prickly issues that constantly erupt in this situation. The water issue for New York and all other cities is in itself a microcosm of the reach of cities—and the sometimes negative impact of the sheer existence of cities on surrounding regions many miles distant from a city's core.

Water has always been a touchy issue. Who controls the water coursing from the interior uplands in progressively widening rivers as it nears the sea has always been a source of contention; people downstream are constantly worried about water being diverted farther upstream. The rise of cities is in no small measure the story of the successful political control of waterways. The political unification of Egypt—of the primordially separate fertile delta near the Mediterranean (Lower Egypt) with the four hundred miles or so of broad and fertile Nile Valley (Upper Egypt, stretching from Cairo south to the first cataracts of Aswan)—was accomplished, legend has it, when King Narmer seized control of this vast expanse of river and land.

Integrated control of waterways (including the periodic flooding of major rivers such as the Nile) is perhaps *the* essential ingredient of the growth of the early political states that arose after people developed a settled existence based on their early successes with agriculture. But why *cities?* After all, as recently as the 1970s, a satellite photograph of the Nile delta showed an astonishingly regular (though presumably not preplanned) distribution of towns and hamlets dotting the landscape. Only Cairo at the southern end of the delta; and Alexandria, the seaport on the Mediterranean; and a few other coastal sites were bona fide cities. The rest remained smaller concentrations of humanity.

So why cities? Think of another major side effect of farming, of taking the production of food directly into our own human hands: from the beginning, farming was for the most part a success. Local populations, no longer limited in numbers to the carrying capacity of the landscape around them (as had been the case for our hunter-gatherer progenitors), began growing almost immediately. Not everyone was needed to labor in the fields—and indeed, other needs prompted by settled existence quickly surfaced. In short, developing agricultural societies experienced an almost inevitable division of labor. Some people worked the fields; others made shoes, wove fabrics, or made beer and bread. Some worked metals into tools and weapons; some tended livestock; others made pots. Items were bartered, and later, with the invention of a monetary system controlled by the central government, people were paid in cash and bought their food, their shoes, their tools—all their necessities and more—with cash (or credit). The rudimentary division of labor typical of hunter-gatherers—with men doing the hunting, women and children the vegetable gathering and food preparation—became far more intricate. Some people could lead comfortable lives without thinking for a moment about where their food came from, as long as the merchant down the street kept selling the bread, the meats, the spices and fruits they could buy or trade for.

Again, water is the key, at least for early cities. Rivers were the highways of the ancient world, and although camel caravans could traverse the Silk Road and cross vast stretches of the Sahara, most of the early commerce was along rivers and, eventually, over stretches of large lakes and even seas. Rudimentary trading was known to Neanderthals and other early humans: in some areas, the best material for stone tools, for example, was often hundreds of miles away from where the tools have been found—yet it made its way far from the local outcroppings that were the only known source of supply. The trading habit was stimulated by the division of labor that sprang up with village life, itself a product of the agricultural revolution.

Water was power: for crops to feed the populace, as a source of political power, and as a highway that permitted the maintenance of order through power, and which supplied a means of moving people around and a means of exchanging material goods. Political power was concentrated in the hands of a few—or of a single king, emperor, or Pharaoh-God—while the economic power had to be concentrated in relatively few locales. The places best suited to meeting all these needs were generally very close to water. These were the places where population, material wealth, and political power tended to congregate and accumulate. They were places from which power radiated, and to which people could retreat (often behind fortresslike walls) when the enemy penetrated the outer defenses and drew close. These were the earliest cities.

CITIES AND THE NATURAL WORLD

What could be more different than a city like New York and a wild place like the Grand Canyon? New York is a paved-over, concrete, steel-and-macadam-ridden chunk of real estate. The Grand Canyon, in contrast, is a mile-deep chasm cut through the layers of the Colorado Plateau by the Colorado River, a gleaming ribbon to the eyes of the visitor gazing over the edge of the precipice. In many ways, cities like New York are the apotheosis of environmental destruction. What was once an ice field, then a tundra, then a forest is now a jumble of buildings and roadways. Places like the Grand Canyon, in contrast, survive either because their awesome beauty or remoteness—or in the case of the Grand Canyon itself, its sheer physical structure—preclude mass occupation by human beings. The Grand Canyon, much like New York, in the old saying, is "a nice place to visit, but you wouldn't want to live there."

True, people flock by the thousands in good weather to visit the Grand Canyon. And some intrepid souls walk or ride mules or horses down to the Colorado River and back up again, or they sally forth in rafts to brave the rapids of the river. You do not have to look far to see the "hand of man" imprinted all over the place. And on the other side of the ledger, trees actually do grow in Brooklyn—and in Manhattan, too. The garden plots, roof gardens, and extensive parks of New York everywhere refresh the urban environment, deliberately recalling the natural woodlots and fields that not long ago dominated Manhattan Island. But if there is people-induced degradation in almost all wild places, and if there are parks and gardens in even the most crowded of

Figure 5. Roof garden. Such gardens add to the overall plant diversity of New York City, and roof-garden diversity is as great as that of ground-level gardens. Plants grown in the city range from common species to specialty plants, such as types also found in alpine gardens. Photograph by Sidney Horenstein.

cities, such as New York, on the whole the differences are nonetheless glaring: New York is human-made, the Grand Canyon is of nature.

Nature has been taking it on the chin ever since people came along—people with increasingly sophisticated cultural techniques for hunting and otherwise taking from the natural world what they need in order to live. And when agriculture came along, "fageddaboudit!" as New Yorkers are prone to say. Tilling the fields, and keeping all but the planted crops and invited stock animals out, is a direct declaration of war on the very existence of the natural world. Never before had humans so passionately seen themselves as conquerors of nature—as beings perhaps made in God's image but, in any case, separate and distinct from the beasts of the field and all the rest of the natural world. Sportscasters on New York AM radio routinely refer to the Hackensack

Meadowlands (across the Hudson River in New Jersey—home to both the "New York" Giants and Jets) as "the Swamp." New Yorkers and Jerseyites for the most part continue to see swamps as places where bad things happen, such as disease, as in malaria ("bad air"), or Mafia body-dumping. They especially see them as wasted space, places where a little landfill will go far toward making room for new condos, shopping malls, and sports arenas.

Yet the Jersey Meadowlands, washed twice a day by oceanic tides, are historical nursery grounds for many species of marine fishes and invertebrates, including significant components of the Atlantic fisheries that, until now, we could expect to find at the fish market or restaurant. Tidal wetlands also provide a buffer between oceanic storms and the land, as well as a source of flood protection. Loss of wetlands up and down the coast—and all over the world—explains in part why 90 percent of the major commercial fisheries of the oceans are severely depleted and, in some cases, faced with imminent collapse. There is a price to not knowing where your food comes from—other than from "the store."

Here's the problem: the invention of agriculture, followed quickly by the rise of cities and the flood of humanity that has yet to reach its peak, has put tremendous pressure on the lands and seas of the earth. At first glance, this may seem fine, for city dwellers in general *like* cities. (New Yorkers are notorious for not being able to imagine living *anywhere* else!) We have been on an extended honeymoon away from nature ever since agriculture came to inform the lives of the majority of the earth's people. We see ourselves as outside nature, which is neither wrong nor all bad since, as we have already noted, agriculture really does take people not just away from but truly *outside,* the living world.

Yet even though we are outside nature, we humans survive by wallowing in the wealth of nature's biodiversity. United Nations statistics suggest that people the world over—in cities but also in rural areas, in developed as well as in so-called third world nations—rely on a minimum of forty thousand species of plants, animals, fungi, and microbes on a daily basis just to keep going. We scour the seas for fish; we cut trees for fuel and building materials. We use plants for medicinal purposes as well as for food. Wild species play an important role in the ongoing search for medicines. In the world of modern biochemistry, we can synthesize most of the drugs we use to combat disease. But we can't make what we don't know about. So the major pharmaceutical companies are keenly interested in exploring the natural pharmacopoeia of plants (and even animals, such as coral-reef invertebrates) whose natural chemical products have already been discovered by indigenous peoples to have beneficial medicinal effects, or which, through direct experimentation, may prove to be the case. That the plants and animals of the world are rapidly disappearing is not only poignant but also a cause of great alarm among those who would document the vast untapped treasure troves of molecular pharmacology trapped in wild plants and animals before they all disappear.

Even agriculture itself, the very thing that plucked us from the arms of Mother Nature in the first place, is still tied to Mother's apron strings. Since all domesticated plants and animals came from wild species, the more we narrow the genetic variation in our domestic crops, the more urgent it is to locate the ancestors of domestic corn, wheat, apples, and oranges in the wild. We need ancestral genotypes so that we may study and even extract the original genetic materials, perhaps to help reinvigorate domestic lineages and to convey their resistance to disease and climatic events that has been lost over the ages as we have single-mindedly bred our domestic crops for high crop yields.

Beyond seeking food, shelter, and medicines, we rely as well on what is left of natural ecosystems to regulate water cycles and to produce oxygen and fix nitrogen, all essential ingredients of continued human existence. After all, no matter how civilized we take ourselves to be, we are very much still animals. We need oxygen; we need water; we need nitrogen. Nitrogen is an essential element in all proteins; no animal, certainly including human beings, is able to extract nitrogen directly from the environment and incorporate it into his or her body chemistry, despite the fact that the atmosphere is 78 percent nitrogen. Only some bacteria and fungi—often in association with the roots of leguminous plants such as the members of the pea family or the clovers—can fix nitrogen in plants and thus make it available to animals (including us) when they eat those plants or eat the flesh of other animals who have eaten such plants.

Once again we confront the subject of water. Seemingly ubiquitous, it is an increasingly rare and precious commodity for fully one-third of the world's population, who have no ready, dependable access to freshwater supplies safe enough to drink. Water fills the oceans, but *fresh*water, the kind we humans use, must constantly be replenished. Water evaporates from standing bodies like lakes and ponds and, especially, the oceans, and it falls as rain, sleet, hail, and snow. Water runs down the hills in rivulets, merging into streams and then mighty rivers before it once again reaches the sea.

Water is trapped—at least temporarily—in lakes, ponds, and bogs; it saturates the ground, forming the water table. Part of the problem of access to water these days is that much of the water is either diverted, used up, or so badly polluted it can't be used. But a large part of the problem is destruction of the world's ecosystems: water is recycled through plant life. Plants take in water and carbon dioxide, which they transform, through photosynthesis, into oxygen, sugars, and water. Plants hold soil with their roots, retarding erosion, and they transpire—which, in the tropics, is the usual cause of the daily afternoon thundershowers during the rainy season. Plants also filter harmful chemicals from the air; nowhere is this most evident than in cities, in which air is measurably cleaner where trees are allowed to grow.

Paradoxically, then, people continue to need the natural world, and all the more so as our numbers grow and more and more of us lead the life of the inner-city resident. As cities have grown, it has become harder and harder to see that this is so, but it is indeed the case. Others have argued that, no matter how wedded to city life we may be, the lush greens of the countryside are part of our psychic makeup, even to the point of being somehow part of our genetic makeup (this is the notion of biophilia, developed by Harvard biologist E. O. Wilson).

The human population has skyrocketed since we took life into our own hands with the invention of agriculture. As our numbers have expanded, so have the acres of farmland we have put under cultivation. We have already exploited virtually all of the earth's natural habitats in our quest for resources. Human success, as expressed by our sheer growth in numbers, has come at the expense of the natural world—so much so that all the world's species (there are at least 10 million species) living in all the world's ecosystems are suddenly faced with extinction. Indeed, some are already gone, and an estimated thirty thousand species a year are currently being forever lost.

This is the great biodiversity crisis—the period of human-induced environmental change that has begun to drive entire species extinct. Some biologists have called this loss of species through the depletion of ecosystems the Sixth Extinction, since it resembles so closely the events of the past that also resulted in the net loss of millions of species. The last great mass extinction occurred 65 million years ago, when a huge asteroid struck the earth, creating ecological havoc and consigning many kinds of animals and plants, on land and in the sea, to oblivion. Among them were the dinosaurs—who had managed to survive for 150 million years before this cataclysm took them away. And while it may be happy news that, with the passing of the dinosaurs, we mammals began our ascendancy, consider this: it took the earth at least 5 million years to recover to the point where life had regained a semblance of normalcy.

Cities in many ways epitomize environmental destruction. Nothing else on earth comes as close as a city like New York to total environmental subjugation. To build a city, you not only cut down trees and plow under grasslands, but you also fill in entire lakes, remove entire hillsides, and displace huge quantities of earth and rock to plant the roots of tall buildings and to construct the tunnels where water, waste, subway lines, steam pipes, and electrical lines snake their way around as literal infrastructure.

Furthermore, because of the city's insatiable thirst for water and its need for food and other products, its reach extends far beyond its footprint, into the surrounding countryside. Nowadays, in the global economy, the reach of New York City is all the way around the world as traded goods depart from and arrive daily in New York's harbor. And perhaps even more critically, places like New York serve as financial hubs for the entire global trade apparatus, which exceeds $1 trillion dollars in trade *every day!* It is as if cities, blinded by their own needs and desires, and running on sheer need to keep the business flowing, are the ultimate source for much of the world's environmental destruction.

The picture is grim. But there is another side to it: if cities in themselves—and through their mighty and far-flung effects on distant lands—are the quintessence of environmental destruction, they also hold out our best chance to strike a balance, to conserve enough of the world's natural systems that human life will persist along with a goodly chunk of our biodiversity, whatever remains of the world's species and ecosystems.

Once again, there are local, regional, and international aspects to the *good* that cities can and do present to the natural systems of the world. Locally, through conservation of remnant wild areas, through preservation of unique animals and plants, and especially through the determined establishment of parks and gardens, cities actively recognize the needs of their inhabitants to keep in touch with the wilder, more natural parts of their earthly heritage. But even as cities reach their destructive tentacles far beyond their own physical and political limits, they also have the very strong potential to reverse, or at least slow down and alleviate, the tide of destruction now engulfing the entire planet. Most of the world's great cultural institutions—including institutions of higher learning such as universities, along with research institutes such as museums and zoos—are located in cities. The great banking and political forces are there, as are most of the media. If solutions are to be found to the current problems facing all of life on earth, human and otherwise, they are most likely to be found, developed, and promulgated in the world's great cities.

That, in a nutshell, is the yin and yang of cities with respect to the "forest primeval," which Henry Wadsworth Longfellow mentioned in his epic poem *Evangeline*. Cities need healthy chunks of the world's ecosystems to persist if they themselves are to persist; yet cities, like parasites, grow and prosper by local destruction of these very ecosystems. At once wonderful and terrible, cities offer the very best and the very worst of human existence, especially when we think of what the world was like just ten thousand years ago.

New York City is a youngster as far as cities go. Damascus in Syria is said to be the oldest of the world's continually occupied cities. Cairo was founded by the time of King Narmer, 3100 B.C. But whatever their age, cities represent that most human of impulses to associate in large numbers, protected from the usual limits nature imposes on life. Some Old World cities, such as Cairo, stand on ground occupied by prehistoric people so long ago that the surrounding deserts were then meadows filled with waterbirds and papyrus. Cities in the New World, such as Los Angeles and São Paulo, sprang to life in the wake of the European conquerors who followed Columbus. And cities like Beijing and New Delhi, with long imperial histories, have traded their traditions of rarified nobility for the roaring economic power that drives a national economy and concentrates impossibly large numbers of hopeful workers in a largely unplanned megalopolis.

Whatever the history or location, all megapolitan centers face universal challenges to survival. A brief review of their development and their responses to such needs as freshwater, clean air, adequate housing, and rubbish disposal reveals a pattern of underlying parallels in the timing and nature of solutions, even though local conditions or practices reflect unique problems and innovations. New York City and Los Angeles can breathe easy when they compare their air quality to places like Cairo and Beijing, which are listed in the top ten for poor air quality among the world's biggest cities. Yet, when we recall New York City's past, when it was mired in horse dung and blackened with coal smoke, it's no surprise to find planned reforms and environmental controls in less developed cities as they, too, grow up and catch up with places like New York or Singapore.

A 2002 World Bank study calculates a world urban average of 60 micrograms of particulate matter—soot, ash, dust, and so on—per cubic meter of air. Cairo, among the top five worst-affected cities in the world, comes in with levels above 130 micrograms per cubic meter. By comparison, the residents of New York and Los Angeles inhale 22 and 36 micrograms of particulate matter, respectively, in each cubic meter of air. It has been said that living in Cairo is like smoking a pack of cigarettes a day, but there are signs that air improvement efforts are working. To remedy dangerously high levels of lead in the air and reduce exhaust, the government runs a free program to get some of the city's 1.5 million drivers to convert their cars to natural gas and underwrites the purchase of this fuel, so that it sells at the same price as gasoline. Around fifty-five thousand vehicles have been converted, which seems to be helping, since a 2004 pilot survey shows a significant decrease in many air pollutants. In addition, the Egyptian government has moved to control both agricultural burning and industrial sources of pollution such as potteries and foundries.

Beijing's best efforts to reduce pollution remain hazy at best. While considerable green efforts born on the eve of the 2008 summer Olympics continue to thrive, the population of Beijing now exceeds 17.5 million, and coal continues to supply 40 percent of its energy consumption. Meanwhile, congestion remains a challenge, with a thousand new vehicles registered daily. The usual toxic blend of domestic, industrial, and agricultural waste effluent and runoff renders much of Beijing's water unpotable. Of twenty-one reservoirs serving the city, several have dried up as a result of the progressive desertification of the countryside and owing to upstream degradations, diversions, and overuse. Water in the city's fourth-largest reservoir is now fit only for irrigation, and nearly half of Beijing's sightseeing lakes and ponds are too polluted to use for irrigating their surrounding parks. Beijing is currently experiencing severe water shortages, and the Yangtze River diversion has just recently begun sending water to Beijing.

In the New World, São Paulo is the earth's second-largest city, a megalopolis of 10 million within the larger metropolitan extent of Greater São Paulo encompassing a population of over 19 million Brazilians. Many live informally housed in favelas, the squatter settlements that sprout up without basic plumbing, sanitation, and electricity, which are unaffordable to a poor population pouring in from the countryside. To cite a single example, consider the impact of the favelas on the Guarapiranga reservoir located in the southwest of the metropolitan region of Greater São Paolo. The reservoir supplies 3 million inhabitants of metropolitan São Paolo. Mountains of uncollected garbage plus solid and contaminated waste previously blocked the natural drainage system, while liquid wastes drained into the reservoir, polluting the entire Guarapiranga water basin. In 1993 the city and the World Bank joined in funding a cleanup program that has shown positive results: fifty-two favelas in the Guarapiranga water basin gained infrastructure, recreation areas, and new housing units; nearly seven thousand families have seen such improvements, and a program of public information and environmental education helps residents learn how to protect their own environment and the reservoir.

For the 13 million people of Cairo, there is no water shortage on the banks of the Nile, which meets the city's entire freshwater demand, but the accumulated discharge into the Nile's 660-mile channel south of the Aswan Dam leaves it heavily polluted with wastewater. And while Cairo's drinking water is generally rated safe, in squatter neighborhoods such as the City of the Dead, which houses many thousands of living families in the cemetery's aboveground tombs, only public fountains are available. While official estimates calculate that 90 percent of residents have piped-in drinking water, the true figure is probably much lower, and long lines prevail at many public taps. Environmental studies show that as much as 80 percent of Egypt's industrial effluent is discharged into the Nile; at least half of the country's industry is located in the capital.

Yet the Egyptian Environmental Agency, established in 1997, reports progress in controlling industrial pollution. By 2006 the agency reported three consecutive years of improving water quality and the termination of industrial effluent from eighty-three establishments.

When we consider that, until recently, filthy conditions prevailed in the Hudson and East Rivers of New York City, that urban degradation in Westchester County polluted the Croton watershed north of New York City, and that New York City drains water from the Catskills and other outlying regions, our example of New York (see chapter 5 for a fuller account) holds promise for developing cities, even as they may supply some innovative solutions of their own.

In this regard, Singapore sets a standard that offers hope to the rest of the world. Considering the entire nation, with its population of 4.35 million, as the fully urbanized modern city-state it truly is, here we see one of the most densely populated places in the world. Yet a rigorous system of monitoring, regulation, and relocation of industrial sites, and vigilant pollution control, gives the nation access to approximately 1.3 million cubic meters of potable water a day. While the water quality fully meets World Health Organization requirements, Singapore is working to meet even higher criteria.

Garbage is another universal urban dilemma, but wealthy, first-world cities generate far more waste than poorer ones. Cities in the United States can reach waste-generation rates of over 1.2 kilograms per person daily, while the residents of some African cities may generate as little as 200 grams a day, most of it organic, in contrast to the high percentage of glass, plastic, and metal packaging associated with processed and packaged first-world goods. Since waste generated in developing countries generally contains about three times the percentage of organic materials found in industrialized countries, it tends to be denser and more humid, making it less suitable for compacting and landfill. In fact, the hilly location and narrow, ill-paved roads of places such as Mexico City and São Paulo make access for garbage trucks difficult anyway, causing half of Mexico City's garbage trucks in 2002 to be out of commission and in need of repair.

São Paulo's fourteen-thousand-ton daily discard equals the weight of the leaning Tower of Pisa, but the city's warm temperatures and humid climate make solid compacting techniques that are common in the United States unworkable here. Instead, organic trash is sorted out to be laid in windrows for composting, while *catadores,* the trash pickers of São Paolo, improve their lot by earning a better-than-average laborer's wage as they help the environment by scavenging and recycling most of the solid waste. Similarly, the trash pickers of Cairo—the *zabbaleen*—earn about triple the minimum wage there, where a pair of workers with a donkey cart may service up to 350 households a day. Little goes to waste as they collect organic waste for pig food, and human and pig excrement for agricultural fertilizer, and they pass along the scrap metal, plastic, paper, and glass they collect to be recycled.

India's capital city, New Delhi, announced in April 2007 a comprehensive plan to tackle its slum problem directly by redeveloping informal housing, razing unsafe dwellings, and relocating as many as 2 million poor tenants into new, high-rise buildings. The city, which now numbers 15 million residents, annually absorbs about a half million refugees from the poverty-stricken countryside. But construction laws have not kept pace, and since 1950 all building has been regulated by a single municipal board, the Delhi Development Authority; for the first time, tall construction will be zoned for all but protected historic quarters, and private development will be allowed and encouraged.

The present situation in New Delhi combines all the ills of unscheduled housing: as much as 60 percent of the city's population lives in illegal buildings that have no official existence and, therefore, "no safe water supply, no legal electricity system and no proper sewers," as reported by the *New York Times* in the April 13, 2007, article "A Plan to Tame the Architectural Chaos of India's Capital," by Amelia Gentleman. The resultant tangle of electrical wires snaking power from pylons frequently sparks fires that are hard to extinguish as they race along the narrow lanes of unplanned settlements with no formal streets. Moreover, the unscheduled and underserved demand for water sorely taxes Delhi's system, so water flows into much of the city at a trickle, for just a few hours a day. Power outages are common too, and uncollected garbage chokes the streets. As the city government contemplates a predicted expansion to 23 million residents by 2021, it hopes to improve conditions by building upward in a city that until now has mostly limited housing to three-story buildings. Delhi's city planners hope to achieve an expanded water system, broad new avenues, verdant parklands, and many new apartment buildings by expanding upward.

As the home of both Hollywood and Disneyland, the Los Angeles–Long Beach–Santa Ana megapolis is the "dream factory," and it also may well be the world's most dreamed-of city. With a documented population of close to 12 million spread out over nearly two thousand square miles, LA is home to over a quarter of California's entire population, making it the second-largest city in the nation. And while the West enjoys a reputation for wide-open spaces, there is a severe housing shortage in Los Angeles. The *Washington Post* reported 2005 U.S. Census Bureau figures indicating that "ten municipalities in the nation average more than four people per household—and nine of them are in greater Los Angeles," mostly in older neighborhoods of tract houses, where many garages are turned into illegal apartments.

The Los Angeles Department of Water and Power claims to be the nation's largest municipal utility, responsible for both water and power services to the city's 3.9 million residents. In March 2007, the utility announced plans to increase its renewable energy supply to 20 percent by the year 2010, starting with the purchase of energy from several small hydroelectricity-generating facilities in the Pacific Northwest. In November 2006, the city also made a landmark decision to approve the San Pedro Bay Ports Clean Air Action Plan, called the most comprehensive strategy to cut air pollution and reduce health risks ever produced for a global seaport complex. The plan calls for the phasing out of high-polluting diesel trucks servicing posts, stringent goals for air quality improvements, and technological advances to reduce greenhouse gases in this highly polluted American city.

If some of the greatest challenges to a civilized way of life may be found in today's largest urban centers, it's also where we must seek the answers to questions about humanity's future well-being. The very impulses that bring people together in metropolitan settings hold the promise of a cleaner and healthier future. Cities also hold the key to the preservation and reconstitution of the shrinking, truly wild places that still survive on the planet. In the ensuing chapters, as we survey the challenges and solutions of America's leading first-world metropolis, New York City, we will see in microcosmic detail how a small foothold in the New World grew throughout a history that was neither fully planned nor untroubled, and how the city arrived at its twenty-first-century position as an exemplar of metropolitan greatness.

CHAPTER TWO

Forest Primeval

Think of New York City, and you think first of buildings and streets—of steel, glass, building stones, pavement, and of course, concrete. Everywhere. Especially when you think of Manhattan's closely spaced streets lined with skyscrapers, an image of unremitting human construction leaps to mind.

To be sure, there are trees and even parks—some of them fabulous. There are more than half a million trees lining the streets of New York, and over seventeen hundred municipal, state, and federal parks and recreational areas in New York. Manhattan's Central Park, although not the oldest or the largest public park in the United States, is still justifiably the most famous park in the country, if not in the entire world.

BUILDING STONES

A treat awaits any geologist walking the streets of Manhattan. It has perhaps the greatest collection of rocks from around the world—rocks shaped and used in the facades, lobbies, and restrooms of buildings; in the sidewalks and curbstones of the streets; and, yes, in all types of walls. The first settlers, the Dutch, mostly used available material from the local forests, bedrock outcrops, and loose glacial debris as building stones and for the erection of stone walls. Just as their water supply was self-sustaining during early colonization within the confines of the general boundaries of New Amsterdam, so it was with natural building materials. As the city grew, the need for these materials expanded; but the rock outcrops had been chiseled away, and the city had grown over available rock exposures. Soon, these factors led to the search for new material, which took the form of schist in the East Side cliffs adjacent to what is now Tudor City, opposite the United Nations; the marble deposits of the Inwood section of Manhattan, to the north, as well as of what is now the Spuyten Duyvil section of the Bronx; and the tough doleritic rock of the Palisades and the brownstone (sandstone) deposits of Newark and Little Falls in New Jersey, Rockland County in New York, and the Central Valley of Connecticut and southern Massachusetts. Some stone was local enough to transport by cart, while the more distant quarries were all accessible by river vessel.

Coastal New England provided a great variety of granites shipped by stone sloops. That trade expanded greatly with the coming of the railroads, allowing the shipment of stone from greater distances, such as the great limestone deposits of Indiana and a variety of stones from elsewhere in the West. Eventually, the whole world began to export building stones by seagoing vessel and, on occasion, by plane, just as foodstuffs are. Today, no stone-producing locale in the world is inaccessible for architects choosing the stone that meets their needs for color and properties. At the Sony Building (Fifty-Seventh Street and Madison), you can see the stone products called Stony Creek Classic Granite, from Branford, Connecticut; on the facades of Rockefeller Center, Indiana Limestone; on the exterior sculptures of the Custom House (on Bowling Green), Tennessee Marble; and at the New York Public Library, interior walls of marble from Greece.   *SH*

Figure 6. View of the east side of Manhattan, centered at Fifty-Second Street, from Roosevelt Island across the East River. The rectangular grid plan, mapped out between 1807 and 1811, is reflected in building shapes. Photograph by Sidney Horenstein.

Surely, here one finds respite from the hurly-burly concrete jungle; and just maybe all those trees, grassy fields, ponds, and rills in Central Park reflect the survival of a bit of Manhattan's ecological past—a spot where the developer's eye was deflected away.

But we'd be wrong to mistake Central Park for a surviving remnant of the old prehuman settlement ecosystem, a bit of the forest primeval. In fact, as Sara Cedar Miller tells us in her book *Central Park, an American Masterpiece,* not one whit (save the many rock outcrops) of Central Park is wholly "natural." The ponds are laid out by design, their shorelines human-made and their plants carefully selected and cultivated. True, many of the trees are native, but the character of the woodlands is unlike the original stands of trees before humans arrived in numbers

Figure 7. Central Park West at Seventy-Fourth Street. The Central Park landscape and the "city" landscape present a study in contrast. Photograph by Sidney Horenstein.

and began the uptown march of development. Even many of Central Park's rocky outcroppings have been modified to yield the shapes we find today. Central Park is a landscaping triumph, a true work of art and no remnant of the forest primeval.

But for all our human impact on nature, the complete and unremitting impact of urban transformation upon even so circumscribed a place as Manhattan Island is itself an illusion: and with that thought we find ourselves standing in Inwood Hill Park at Manhattan's northwestern corner. We see here a hilly strip of land bounded on the west by the Hudson River (including, to be sure, railroad tracks), and on the north and east by the Harlem River and the U.S. Ship Canal. (The canal, as we shall see in chapter 3, was constructed to allow a more navigable connection between the Hudson and Harlem Rivers, so that ships coming down from Albany could reach the South Street Seaport without having to swing down around Manhattan's southern tip.) Inwood Hill

Figure 8. A waterfall in the northwestern portion of Central Park, part of the area designed to emulate New York's Adirondack Mountains. Here, the original natural stream was dammed to create the waterfall, but today most of the water in the stream comes from the city's water supply system. Photograph by Sidney Horenstein.

Park strongly resembles the natural forest—if not utterly pristine, it is at least a close approximation.

In 1918, seeking chestnut trees resistant to the blight that attacks their bark, Arthur Graves, who was then a professor at Yale's School of Forestry and later a curator at the Brooklyn Botanic Garden from 1921 to 1947, found what he was looking for at Inwood. He was surprised to discover within the confines of New York City that "a wild, wooded section of this sort still existed. Fallen trees, dead decaying branches, weeds and climbing plants, together with the vigorous growth of tall stately tulips and oaks, and an understory of spicebush, all combined to produce an effect of natural wildness that was far remote from the general conception of what is meant by New York City."

Figure 9. Inwood Hill Park schist. Landslides, used as rock shelters by Native Americans, occurred after deglaciation, when the frost cycle was more intense. The exact time of the landslide (or landslides) here has not been determined, but it occurred before Native Americans occupied Manhattan Island. Many artifacts have been found here, some of which are now housed at Dyckman Farm House at 204th Street and Broadway; others are in the collections of the American Museum of Natural History. Photograph by Sidney Horenstein.

Inwood Hill Park still boasts a wonderful diversity of plants: over 250 species have been identified there. Of course, some of these are alien species that have invaded from elsewhere in North America, as well as from Europe and other far-flung places, but it is one of the few places in New York City where you can still find wild Dutchman's breeches (*Dicentra cucullaria*), a dainty corsage of wildflowers with yellow tips, a member of the poppy family. The park has the greatest diversity of habitats in Manhattan: an oak-hickory forest growing mainly in the sloping cleft between, and on the sides of, the two hilly prongs—a valley long-ago dubbed "the Clove." To this day, the Clove sports the stately tulip trees

Figure 10. A black eastern gray squirrel (*Sciurus carolinensis*) posed on the stump of a black willow. This adaptable tree-dwelling species constructs large dens on thick tree branches or within hollow tree trunks. In New York City, there are large populations of melanistic forms, perhaps a result of the lack of predators. Although once confined to the eastern United States, the species has been introduced in parts of the western states, as well as in Ireland, Britain, and Italy. Photograph by Sidney Horenstein.

and ancient oaks, the underbrush of spicebush and witch hazel, and the carpets of wildflowers that so surprised and delighted Arthur Graves. Wild turkeys, red-tailed hawks, and many other resident and summer-breeding migrant bird species are regulars there, while a nesting structure established in 2002 to encourage bald eagles' breeding produced a number of young until it was removed in 2009. The forest floor is home to eastern voles and red-bellied salamanders. Higher in the canopy, nocturnal southern flying squirrels glide from tree to tree. Eastern gray squirrels, unsurprisingly, are ubiquitous. Opossums (relatively recent migrants from the southern United States, and thus not members in

Figure II. Canada geese (*Branta canadensis*), which often feed on mowed lawns in parks and golf courses. Widespread in the United States, the populations declined early in the twentieth century as a result of hunting and loss of habitat. Today, many populations have grown substantially and, in places, are considered pests because of their droppings and confrontational behavior. Some Canada geese populations no longer migrate. Photograph by Sidney Horenstein.

good standing of New York's primeval forest), raccoons, and (until recently, at least) foxes are sometimes to be found—and, given the recent appearances of coyotes in both Central Park and the Bronx's huge Van Cortlandt Park, it's no surprise that a few coyote transients have already been spotted at Inwood. The meadow features grasses and native flowers, cottontail rabbits, meadow voles, white-footed deer mice, and many species of butterflies.

Then there is the salt marsh fringing the Harlem River, with its cordgrass and bulrushes—plants that can tolerate both salt and tidal changes in water height. Fish, mollusks, and crustaceans live in these marshes, which serve, like tidal marshes the world over, as important

Figure 12. One of the last wetlands along Manhattan's shoreline, in Inwood Hill Park. The shorelines of Manhattan, as well as of all the other boroughs, were endowed with abundant wetlands, most of which succumbed to a variety of landfill projects. Even here, the wetlands have developed on landfill-altered shoreline. Photograph by Sidney Horenstein.

nurseries for the world's fisheries. Ducks and geese, herons and egrets, all are readily apparent according to season. If, ecologically speaking, Inwood Hill Park is not exactly like stepping back into precolonial times, it is clearly the next best thing to it.

It was the "Father of Greater New York," one Andrew Haswell Green, who in 1895 first proposed setting aside Inwood Hill as parkland, just one of his many plans to benefit New York City. He was also involved in the inception of the American Museum of Natural History and served as a trustee there from its founding in 1869 until 1881. In addition, he served as president of the board of education and was influential in seeing Brooklyn, as well as the Bronx, Queens, and Staten

Island, become part of New York City during his tenure as the city comptroller. His hopes for Inwood didn't gain much ground at first, but by 1916, because of archeological finds, Inwood's geological features, its historical associations (one unproven story has it that Peter Minuit's transaction to buy Manhattan actually took place at Inwood), its beauty, and its vistas of the mighty Hudson—and its original wooded character—the city finally agreed and began acquiring the land to create the park.

Not to overromanticize Inwood Hill Park. The original natural features of what is now the park were chipped away by the wealthy estate owners who began settling there rapidly in the 1850s, altering the terrain and adding exotic plants. After the estates were abandoned, many of the buildings were occupied by squatters, and those that remained were finally removed in the 1930s by Robert Moses and the Works Progress Administration. And, in the early part of the twentieth century, Jewish Memorial Hospital and other institutions moved into the park area. Arson, dumping, the creation of trails, and poorly conceived erosion-control projects have all diminished Inwood's primeval integrity. The concrete jungle grew at the expense of the original woodlands, marshes, and springs but never quite managed to obliterate it all, although Inwood does bear the scars of human misuse. Still, Inwood affords us a welcome glimpse into primordial Manhattan—starting with its rocky outcrops that help us understand the structural underpinnings of Manhattan's terrain.

GRIT AND GRAIN

Among the many splendors and technological marvels in the Gottesman Hall of Planet Earth at the American Museum of Natural History is a seismograph, a rotating drum attached to sensors that detect the various sorts of waves generated by earthquakes. A needle scritch-scratches a jagged path in black ink as the drum slowly turns. Mostly what you see is a record of background vibrations—including (as also recorded by the first seismograph in the old Hall of Geology) the passing of subway trains along the tracks under Central Park West.

Every week, the *San Francisco Chronicle* publishes a map of the Bay Area showing the earthquakes of the past seven days. Though most are too weak and localized to attract the attention of an earthquake-blasé citizenry, they nonetheless do show up as a series of sharp jumps on the local seismographs. The *New York Times,* of course, prints no comparable map, since earthquakes the size of those routinely experienced in California are relatively rare on the Eastern Seaboard. Whenever movement along a fault line does happen—as when Ramapo Fault (which runs near the Indian Point nuclear reactor) jumps—that's a story worthy of attention in its own right.

But 'twas not ever thus: Manhattan and its surrounds come from a past every bit as geologically violent as the current geological scene in California. And though that past—looking back over a billion years—may seem at first glance to have nothing much to do with New York as we see it now, the truth is that New York's ancient history has everything to do with why skyscrapers are concentrated in Manhattan's southern tip, with a gap between that and the next big cluster in midtown. The very grain and fabric of New York's rocks, sands, and soils have played a huge role in determining the shape of the city's growth.

Take, for example, Manhattan's regular grid work of avenues running approximately north-south neatly traversed by evenly spaced cross-streets. The simplicity of this roadwork is deceptive. True, below Houston Street—and perhaps especially below Canal Street—the roadways tend to intersect at odd angles, reflecting a layout that was far from a rigid gridiron system. The streets down there are remnants of old thoroughfares and cow paths that bypassed ponds and hills, very much a reflection of the shortest, flattest paths available given the physiography of the southern part of the island.

But north of Houston Street we see a simple linear scheme laid out with few irregularities to disrupt it. (Broadway is by far the best example of an irregularity, taking blocks to angle across the grid, bending northwestward as it proceeds uptown.) You get the feeling looking at any map of modern New York City that, when reason began to prevail and it finally occurred to the city fathers to lay out their streets in a more orderly fashion, they just drew a grid and slapped it down on the map of New York.

But it was by no means as simple as that. The avenues run parallel to the Hudson on the west side of Manhattan, and parallel to the East River along the middle of the Island. But these shorelines are dictated by the ancient grain of the rocks constituting Manhattan's very foundations. And it's the grain—the direction in which the rocks line up more or less in parallel rows running northeast—that is the natural direction for the avenues to run as well. The cross streets cut across this natural, rocky grain—and building them meant blasting through schist, going up stream-worn inclines (as when going west on Seventy-Ninth Street toward Amsterdam and Broadway from Columbus Avenue). Or it meant following natural breaks in the rocks formed by ancient faults, such as where 125th extends all the way to the Hudson, and where New York's westernmost subway line leaves the ground to become momentarily elevated between the hills of Morningside Heights and 137th Street. The 125th Street Fault cuts through the West Side's linear hills, and erosion has done the rest.

This northeast-trending grain of the bedrock—for the most part so worn down now that it is rarely noticed, except in northern Manhattan—was born of violence. California has its San Andreas Fault running near the coast, moving a slice of the Pacific Plate northward and sliding against the North American Plate. Eventually, millions of years down the geological road, that westernmost sliver of California will slam into Alaska, docking there and joining lots of other bits and pieces of landscape—aptly called exotic terranes—that traversed the Pacific and fetched up to form a part of Alaska.

New York's bedrock, too, was built from the interactions of geologic plates, but here the story was one of slamming together and pulling apart, then slamming together and pulling apart again. The earliest

Figure 13. Excavation in bedrock for a building foundation in midtown Manhattan. Here the rock formations are at, or close to, the surface. Photograph by Sidney Horenstein.

part of the story we know of began more than a billion years ago, during the formation of the supercontinent Rodinia. At this time a series of lava flows and some sediments were squeezed and metamorphosed to form New York's oldest rocks, the 1.1-billion-year-old Fordham gneiss, now exposed in the Bronx and on Roosevelt Island. After Rodinia broke apart, leaving ancestral North America, a series of volcanic islands formed, beginning about 565 million years ago. This volcanic arc eventually collided with ancestral North America, folding and metamorphosing the adjacent sea sediments into schist and creating soaring mountains, about 465 million years ago. Later, during the Devonian period, a chunk of the Eastern Seaboard—really Avalonia, a small continent that had separated from Gondwana (Africa)—collided with eastern North America. Long before Columbus, long before the waves of European immigrants and ships laden with African slaves found their way to the New World, Europe and Africa had come together—and had met North America in a head-on collision.

And seemingly as a taste of things to come in the form of migrations of wildlife, peoples, and customs, chunks of Avalonia and Avalonia/Europe stuck to North America, clinging as the Atlantic Ocean opened, then closed, then opened again. Oddly, few equivalent clumps of North America have been identified in Europe or Africa, though some recent maps depict Scotland as having been originally part of North America. Europe apparently gave to North America more than it seemingly took, at least in terms of bedrock and, much later, people and even (as we'll see in some detail) alien species of microbes, fungi, plants, and animals.

The bedrock of Greater New York is metamorphic—chewed up, convoluted masses of old sediments and lavas that were put through the twin grinders of heat and compression when Europe and North America started colliding in a series of spasms starting some 450 million years ago, during the Ordovician period, in the so-called Taconian orogeny (*orogeny* is the process of mountain building). This initial uplift phase involved the collision of North America and the offshore volcanic island arc. Mountains are thrust up when continental plates collide head-on: the best example today is the Himalayan range—still going up, the result of the Indian subcontinent breaking off from Gondwana (specifically, the east side of Africa), crossing the Indian Ocean and slamming into the Asian continent. Continental crust is normally forty kilometers thick; the crust is almost twice that thick in northern India, explaining why the Himalayas are the tallest mountains in the world.

The easiest chunks of Avalonia to spot along the Eastern Seaboard are in eastern Newfoundland, Boston, and points farther south in the Piedmont region of the Carolinas. Trilobites, some of them two feet long, have been found in patches of sedimentary rock of Cambrian age (ca. 500 million years old). These trilobites are nearly identical to those found in Wales and Sweden, and are totally unlike any others of the same age found anywhere else in North America, be it western Newfoundland or the Rocky Mountains. When the supercontinent Pangaea broke apart, part of Avalonia was stuck to North America—while other parts went with England and Sweden.

Figure 14. An example of the intensely folded and glaciated Hartland Formation, Pelham Bay Park, in the Bronx. The formation is a result of several periods of plate collisions. After deep erosion exposed it, this rock was polished and striated by glacial action, revealing the details of the deformed layers. Photograph by Sidney Horenstein.

In the New York region, the evidence is tougher to read. Manhattan's bedrock is made up of metamorphic marbles, schists, and gneisses[1] that have been folded, baked, and crumbled, totally destroying whatever fossils they may once have held. The homegrown, native North

1. Leading to the adage "Da Bronx is gneiss, but Manhattan is full of schist." Gneiss (pronounced "nice") is a so-called high-grade metamorphic rock, typically consisting of bands of different mineral composition, hence different colors. The Fordham gneiss contains a lot of pink feldspars, looks superficially like granite, and has been used extensively as an ornamental building stone; schists are lower-grade metamorphic rocks and often contain a lot of mica. The dark schists and gneisses of the Hartland Formation (formed in part from ancient lava flows) make up much of Manhattan, and these too have been used in buildings, walls, and bridges (especially noticeable in Central Park). The Inwood marble (which extends into southern Westchester County) is a metamorphosed limestone formed initially from limy sediments accumulating on the tropical shores of ancient North America. It, too, has been extensively quarried and used as decorative building stone (e.g., Tuckahoe marble).

Figure 15. Inwood marble in Isham Park, near Seaman Avenue, northern Manhattan, the part of New York City where the marble is most clearly and extensively exposed. Formed in tropical waters, the original horizontal limestone layers were subjected to heat and pressure, which caused them to fold and metamorphose to marble. Photograph by Sidney Horenstein.

American metamorphics are not neatly set apart from the deep-water-derived metamorphics found in the eastern reaches of the Bronx, Westchester, and on up the Eastern Seaboard into New England. This is because the Avalonian rocks have been shoved into, over, and through the North American rocks, making it tough to pinpoint the exact trace of the "suture zone" in New York.

The Taconian orogeny began the closure of the old proto-Atlantic Ocean. The Acadian orogeny later on in the Devonian period (ca. 380 million years ago) made things a bit more concrete, bringing Africa and western Europe closely adjacent to the eastern and northern coasts of North America. This time, a flood of trilobites and other marine invertebrates jumped ship—migrating into our waters from what is now Germany and Morocco, bringing a whole new look to the marine life

Figure 16. Fordham gneiss in the Riverdale section of the Bronx, showing typical light-and-dark banding as a result of mineral segregation. Most Fordham gneiss began as a variety of lava but was then deformed and converted to gneiss by several plate-tectonic events. Photograph by Sidney Horenstein.

that crowded the warm, shallow seas of the North American continental interior. The evidence for this invasion lies west and north of the city, in the foothills of the Poconos and Catskills, and farther west, in marine sediments running across central New York State and as far west as Iowa, as well as up and down the Appalachians. Back then, New York stood virtually on the equator.

The Catskill Mountains were born of an ancient delta of red muds that accumulated rapidly as a vast mountain range rose up in what is now New England. The same type of sediment can be found on the *other* side of that prodigious set of peaks—this sediment too is red, and it too is composed of the muds and sands that eroded off the fast-rising mountain range. For example, this sediment type is now found in Scotland (where it's called old red sandstone) and Norway. What is now

Figure 17. Intensely folded Manhattan schist in Central Park. This rock originated as mud on the deep seafloor as much as 565 million years ago. Heat and pressure from tectonic plate collision 465 million years ago altered and folded the rock, forming schist, and thrust the mass onto the edge of the North American continent. Photograph by Sidney Horenstein.

underneath New York City itself helped form the roots of those mountains.

The final blow to New York's basement rocks came in the so-called Alleghenian orogeny some 320–290 million years ago (during the Pennsylvanian period), the last of the three paroxysms that brought the continents together into the single massive supercontinent of Pangaea (meaning "All Earth"). That was the one responsible for folding the Appalachians into the accordion-pleated terrain we still find them in today, as well as for uplifting the rocks in the New York region. The rocks exposed in Central Park today then lay deeply buried under the still-rugged terrain.

But things were destined to change. By the time the earliest dinosaurs were walking the earth in the Upper Triassic period (some 210 million years ago), a crack appeared in Pangaea, more or less along the original North American–European–African contours. The continents began to pull away from each other, and in due course the crack widened, and chunks of continental crust foundered, giving way to oceanic crust. The early stages of the Mid-Atlantic Ridge appeared, gushing basaltic lava to either side, forming new ocean floor and pushing Europe and Africa to the east, North America to the west, at the slow-seeming rate of an inch to an inch and a half a year. Slow seeming, that is, if you take into account that the process has gone on unabated since the Jurassic period. That crack is now three thousand miles wide when measured between the United States and North Africa.

The first phases of continental breakup brought profound changes to the look and feel of New York. Basins began dropping down, much as they are still doing today in the great East African Rift Valley system. Lava gushed over the land surface, and wedges of hot liquid basalt insinuated themselves between the fast-accumulating layers of sediments deposited in myriad streams and lakes. What is now New York lay under this sinking system of down-dropped basins—and to see the remnants, you have to stand along the shores of the Hudson and gaze across at the great Palisades cliffs of New Jersey, a product of the most massive of those diabase intrusions.[2]

Beyond the Palisades, you are in the lowlands section of the Newark Basin—one source of the rock for all those brownstone buildings in New York and Newark. Out beyond the first valley are a series of ridges—the Watchung Mountains—which are true Jurassic lava flows and are still quarried extensively as traprock or riprap and used extensively in paving material such as concrete.

2. Take a ride over the George Washington Bridge and follow the signs to Interstate Route 80; as you descend the fifteen-degree slope of the back side of the Palisades, look up to your right (better if you are not doing the driving!) and you'll see some sediments sitting on top of the diabase (volcanic rock similar to basalt). Those sediments were baked by the liquid, lavalike basalt as it was injected between the layers, excellent evidence that the Palisades are an injected sill, and not a lava flow that emerged through cracks and simply spread out over the surface of the land.

Figure 18. Columnar joints in the Palisades diabase, in New Jersey, the result of shrinking as this once molten rock cooled below the earth's surface 190 million years ago. Subsequent erosion removed the overlying rock as well as two-thirds of the original thousand-foot thickness of the formation. Photograph by Sidney Horenstein.

Fossil fish and some strange fossilized gliding reptiles have been found in lake deposits scattered along the Newark Basin. They have appeared in places as close to New York as the New Jersey side of the Lincoln Tunnel (featured in the opening segment of the TV program *The Sopranos*), at the abandoned Granton Quarry on Tonnele Avenue in North Bergen, New Jersey, and at the foundation of the main library of Princeton University. Those early dinosaurs were here in the future New York City too, leaving their footprints especially in the muds and silts that are exposed along the Connecticut River Valley. The hills around New Haven are lava flows (and one true sill—an injection of volcanic materials between two layers of sedimentary rock), and the basin of the Connecticut River Valley was actually a part of one giant basin with the Newark Basin—of course including New York City. That changed as later upheavals elevated New York and southeastern Connecticut, erosion took over, and all those red beds, fish, and dinosaurs were dissolved into oblivion, when the bedrock of New York was finally exposed.

But New York–New Jersey's masterpiece of a dinosaur-age fossil has got to be the famous Fort Lee phytosaur still on display on the fourth floor of the American Museum of Natural History. Three geology students were out exploring along the base of the Palisades in 1910, more or less where the George Washington Bridge attaches to New Jersey. (The George Washington Bridge wasn't built until 1931—years after the students made their famous discovery.) Phytosaurs are reptiles that look distinctly like crocodiles, except that their narial openings (a.k.a. noses) were up on bony promontories close to the eyes, instead of at the end of their long snouts as they are in true crocodiles. Neither crocodiles nor dinosaurs, phytosaurs are great examples of the exuberance of "experimental" evolution, evolving at a time when many other forms of now-extinct reptiles, including dinosaurs, of course, but also flying reptiles, plesiosaurs, and ichthyosaurs, populated the planet.

Phytosaurs were thriving in the Triassic-Jurassic lakes all up and down the Eastern Seaboard, lakes that teemed with fish. Lakes that were there because the continents were drifting apart. The bones of

Figure 19. A slab with disarticulated phytosaur bones, found in Fort Lee, New Jersey, shown adjacent to a phytosaur skeleton from North Carolina. The prefix *phyto,* which refers to plants, is a misnomer in this case. When originally described, the animal was thought to be an herbivore, but in fact it was carnivorous. (The skeleton is American Museum of Natural History catalog specimen no. 1.) © American Museum of Natural History/Craig Chesek.

many phytosaurs are black (carbonized)—but not those of the Fort Lee phytosaur, which are said to have gleamed especially bright white when wetted suddenly in a private moment by one of the students. Never mind *how* the discovery was made—the Fort Lee phytosaur is still the only large, well-preserved reptilian remnant of those ancient, rendered-continent basins we've found yet in Greater New York.

So what happened next? We still have something like 200 million years to account for—and here the evidence is, well, inferential, if not downright nonexistent. The Atlantic Coastal Plain nowadays basically starts in New Jersey and runs south (though appendages to the coastline, such as Long Island and southern Staten Island, are also considered part of the coastal plain). Staten Island, part of New York (though maps make it look

Figure 20. Coastal plain sediments exposed in an old clay pit (now destroyed) in southern Staten Island. The sediments were deposited on and adjacent to the shoreline along the edge of the Atlantic Ocean during the Cretaceous period, 100 million years ago. The clay layers have been used to make bricks and terra-cotta. From *Studies of Cretaceous Coniferous Remains from Kreischerville, New York,* by Arthur Hollick and Edward Charles Jeffrey (New York Botanical Garden: 1909).

like it logically belongs to New Jersey), is the first place that rocks—hard old rocks such as serpentine, remnants of ancient volcanic activity—crop out as you travel northward from the Jersey shore, before you get to New York and, after that, New England and its famous rocky coast. But that's because of the effects of glaciers, as we'll see in a moment. Before the glaciers appeared, we have these millions and millions of years, and we can only assume that the New York region, once the very roots of a mighty mountain system, was worn down to a nub, becoming part of the coastal plain. There are dinosaurs in the Cretaceous coastal plain sediments of southern New Jersey,[3] and it's a pretty safe bet that there were Cretaceous dinosaurs ambling along what is now Broadway in Manhattan.

3. The very first dinosaur ever discovered in the United States—a skeleton of a vegetarian, *Hadrosaurus*—came from Haddonfield, New Jersey. The original owner had it on display, and one of the leg bones is said to have served as a doorstop.

But we'll never know, because at some point, probably never to be determined with any precision, what is now New York City, southern Westchester, and southeastern Connecticut was uplifted—presumably slowly and gently, since we do not find the scars that catastrophes tend to leave behind. Erosion set in. Ancient bedrock was exposed. And that's what the glaciers found as they worked their way south and carved out the rest of dear old New York as we know it today.

NEW YORK SANGFROID

We are standing on dry, frozen ground fifteen miles offshore from Battery Park and Staten Island, around where the Ambrose Light tower is permanently anchored in seventy feet of water. That's right: *standing,* for this is eighteen thousand years ago and there is no water here. We are still eighty miles from the shoreline, where the continental shelf gives way to the continental slope. Sea level the world over has dropped precipitously, because glaciers have covered almost half of the earth's land and locked up a lot of its water in the fourth and most recent (and so far the last) of the tremendous surges of continental glacial ice that covered much of the Northern Hemisphere during the last million and a half years.

We are in fact standing on the tundra, home to mosses, lichens, a few flowering plants, and here and there a low-lying shrub. Mastodons, for sure, and most likely wooly rhino, wooly mammoth, caribou, arctic foxes, wolves, and polar bear are among the few mammalian species traversing the frigid plain. The summertime brings plovers, sandpipers, and other birds in to breed, just as in the present-day world they travel considerably farther north in the arctic summertime to mate and nest.

Eighteen thousand years seems like a long time ago, but of course, compared with the vastly greater ages of the geological events that shaped New York's bedrock, eighteen thousand years is a mere eyeblink. And what monumental changes were wrought by these glacial invasions from the north! Can you imagine standing at Ambrose Light back then and

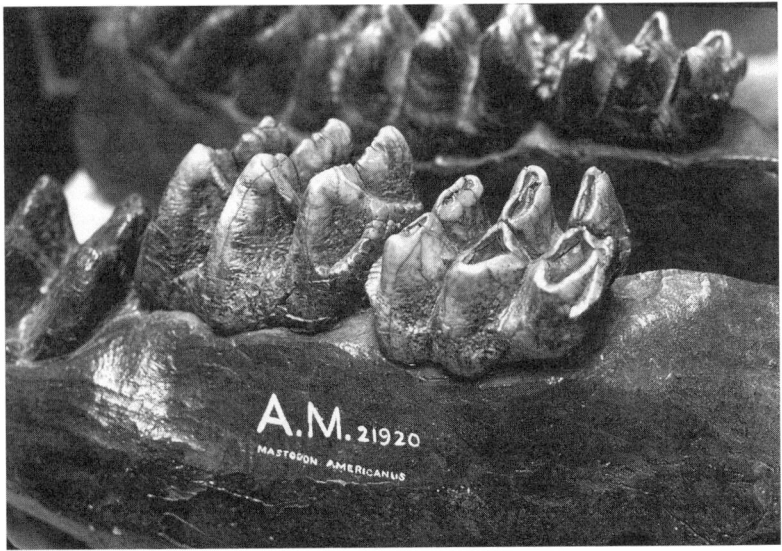

Figure 21. Lower jaw of a mastodon found at Seaman Avenue, near Dyckman Street, in northern Manhattan, now in the collection of the American Museum of Natural History. Mastodons roamed the New York City area after the glaciers receded eighteen thousand years ago. Photograph by Sidney Horenstein.

looking at what is now the New York City skyline—and seeing, instead of the spiky array of Manhattan skyscrapers, the leading edge of an endless white sheet of ice whose thickness reached a maximum of something like three thousand meters? At its southern edge, over what is now Manhattan, the ice was probably as thick as a thousand feet. From frigid tundra and unending ice in the past, to sweltering August days in the present—days when, up until the relatively recent advent of air conditioning, the only relief came from open hydrants and the occasional breeze wafting over the stoops and fire escapes where many a summer night was spent—all that change in only eighteen thousand years! New York's connections with the rest of world—most easily seen by its human links through trade, communications, and continuing influx of peoples from around the world—go back to long before the relatively recent arrival of *Homo sapiens* as the glacier from the last of the four great Ice Ages melted away. As we have seen, hundreds of millions of years ago we were literally connected to both Europe and Africa, until the Atlantic Ocean opened up and the continents drifted apart. But the Ice Age glaciers themselves also provide compelling evidence that we were then, and we remain, physically connected to the rest of the world. For the ice that came to cover New York, shape its present landscape, and create Long Island not only began its creeping growth in the far north but was also triggered by events that happened down south, very near the equator.

What causes glaciers to grow and retreat? And why do ice ages occur only at sporadic moments in the long history of the earth? The earth's tilt and its elliptical orbit means that there are natural cycles regulating the amount of heat the earth receives from the sun. But there are times when, during the course of the year, we are closest to the sun in the winter and farthest in the summer—so those summers are cooler, and the winters warmer, than "normal." (And we all know that climate change can happen so fast that normal itself is difficult to define.)

On top of this regular, periodic fluctuation in solar radiation striking the earth, two and a half million years ago the movement of the oceanic plates finally formed a continuous land connection between North and South America: the Isthmus of Panama was formed. Many years later, when the isthmus was seen as a hindrance to global trade, the Panama Canal was dug (at the cost of an estimated twenty-five thousand human lives and $387 million)—shortening the cargo-ship trip between San Francisco and New York by eight thousand nautical miles and, in a sense, rectifying the strategic problem posed by geological events 2.5 million years earlier.

But there were more immediate effects when the Panamanian isthmus cut off the circulation from the Caribbean into the Pacific, which was the dominant direction of the currents back then. Suddenly, the tropical, westward-bound currents of the Atlantic had no place to go. So like a billiard ball hitting the edge of the table, the currents were deflected—north, it turns out. And the Gulf Stream was born.

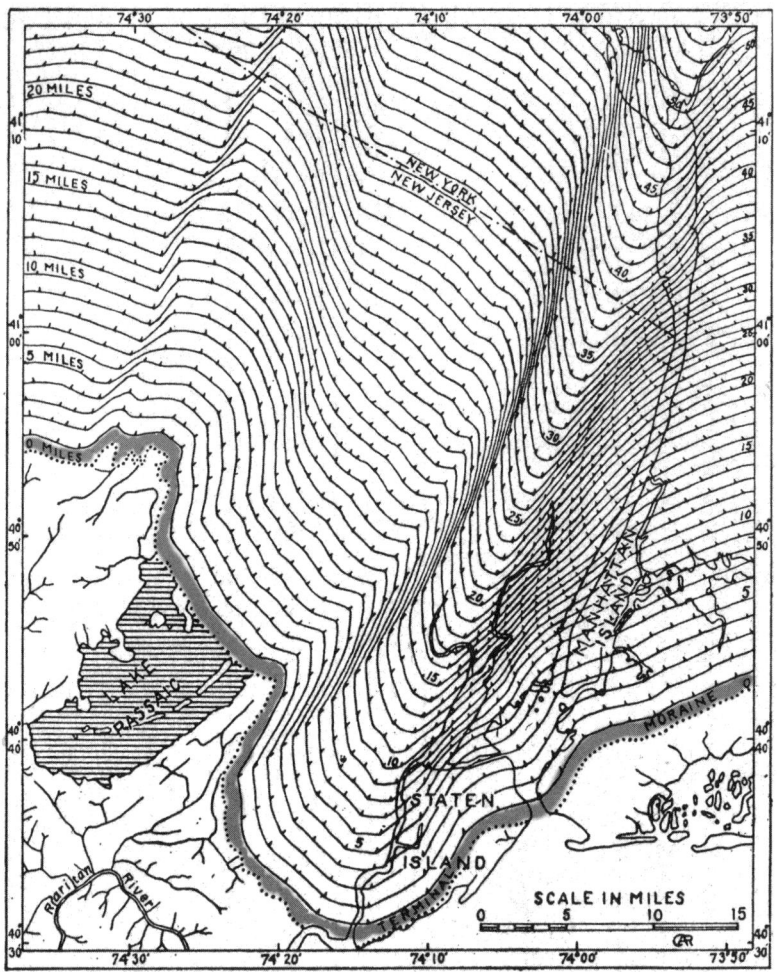

Figure 22. Maximum extent of the continental glacier in the New York region, 21,500 years ago. At that time, sea level was three hundred feet below the present level, and the New York shoreline was one hundred miles away. From *The Geology of New York City and Vicinity,* by Chester A. Reeds. Guide leaflet series no. 56, April 1930. New York: American Museum of Natural History. Courtesy of the American Museum of Natural History.

Figure 23. The Panama Canal, one of the world's great engineering projects. Since 1914, this forty-eight-mile-long waterway has connected the Atlantic and Pacific Oceans, eliminating the need for a long and hazardous trip around the southern tip of South America. The canal is still an important link in international trade. Photograph by Sidney Horenstein.

The Gulf Stream today is the reason why London, Paris, and Amsterdam have mild climates—even milder than New York's weather, although New York lies at the same latitude as Madrid, and London is opposite Belle Isle, Newfoundland. There are palm trees in parts of England—but none at all in maritime Canada. The warm, salty waters of the Gulf Stream cut across the Atlantic and hit European shores, just about at the southern end of the English Channel.

Ironically, these warm, salty waters apparently triggered a global cooling spasm. So ice ages are triggered by a combination of geological events that change oceanic circulation patterns, as well as by the complex astronomical periodicities that change the amount of solar radiation reaching the earth. New York has been connected to the rest of the world seemingly forever.

Four times in the past one and a half million years, a glacier crept south; each time, like a gigantic bulldozer, it pushed a wall of boulders, gravels, sand, and soil before it. Each time that the glacier reached its maximum extent and began melting back, the rocks and sands were left in a long, linear heap—the so-called terminal moraine. Although there is current discussion about the age of the earliest glaciation on Long Island, it is generally thought that the earliest glacier arrived about seventy thousand years ago, during the initial phase of the Wisconsin glaciation; its terminal moraine forms the southern part of Long Island. The tip of Long Island is split in two, evidence that the final glaciation at the end of the Wisconsin (the one that melted away from the New York region only eighteen thousand years ago) did not reach quite as far south as the east end of Long Island. Its terminal moraine contributed to the northern reaches of Long Island and formed Orient Point, the terminal finger of Long Island, just to the north of Montauk Point.

The terminal moraine of the last glacier courses through the boroughs of Queens, Brooklyn, across the Narrows, on across Staten Island, and through New Jersey and points west. Bay Ridge, Forest Hills, and Ridgewood are all familiar neighborhood names—names derived from the very fact that these communities are perched atop that last great glacial terminal moraine. And although Manhattan was covered by ice—and thus was too far north to have the moraine draped over it—from the vicinity of the Metropolitan Museum of Art's medieval collection at the Cloisters, looking eastward, it is possible on a clear day to see the moraine that forms the ridge on the north shore of Long Island.

But plenty of signs of glaciation remain on Manhattan itself. A stroll through Central Park reveals many a boulder perched precariously atop bedrock outcrops. These boulders are generally rounded and show the typical scratch marks that are the signs of the wear and tear of glacial transport and the ravages of fast-moving meltwaters. The very largest of these boulders are chunks of local gneiss and schist, showing that they did not come from afar. However, small glacial erratics (boul-

Figure 24. A glacial erratic relaxing in Central Park. After surviving a trip while embedded in the glacier, this large boulder was used as an ornament by the park's designers. Photograph by Sidney Horenstein.

ders carried from their source area and dropped elsewhere by melting glaciers) are often pieces of Jurassic diabase—the rocks forming the cliffs of the Palisades across the Hudson River. More subtly, but more profoundly, the sculpted, gently rounded outcrops in Manhattan (in Central Park and elsewhere) got their outlines from the relentless grinding of ice and the rocks they dragged (like very coarse sandpaper) over Manhattan's surface—indelible signs of the presence and passage of mountains of ice.[4]

As the last glacier began to melt back, vast lakes formed between the retreating edge of the ice and the prodigious terminal moraine, which

4. In Margaret Mead Green, on the northwest side of the American Museum of Natural History, there is a display slab of Silurian limestone from the Chicago region. It has smooth, deep, rounded, parallel grooves that were carved in the Midwest by the same glaciation that did so much to shape New York. Limestone is softer than gneiss and schist, so the glacial carving is even more dramatic in this imported specimen than any that can be seen on New York's own bedrock.

Figure 25. Glacial grooves, Umpire Rock, in Central Park near Sixty-Fourth Street. This is the best place in the city to see glacial polish, striations, and grooves, as well as the complex structure of the schist itself. Photograph by Sidney Horenstein.

now began to act as an enormous barrier dam. Exposed land was still frozen treeless tundra, and the Atlantic shoreline still lurked way off-shore. Water levels reached so high in some of the lakes trapped behind the moraine that finally the waters began breaking through with force around thirteen thousand years ago. The most spectacular breaching of the terminal moraine was undoubtedly the abrupt flooding event that created the Narrows—that channel between Brooklyn and Staten Island, the main entrance to New York Harbor and now the site of the famous Verrazano-Narrows Bridge spanning the gap between the two boroughs.

From then on, the outflow from meltwater and the newly emerging Hudson River was prodigious, racing across the still-exposed surface of the continental shelf to get to the sea. This extended version of the young Hudson River immediately began digging a deep channel into

Figure 26. Postglacial New York harbor. Before sea level rose, the continental shelf was exposed and the Hudson River flowed to the distant ocean. From "A Geological History of New York Island and Harbor," by J. S. Newberry, *Popular Science Monthly*, 1878.

the soft sediments of the coastal plain—and in a matter of perhaps only a thousand years or so, a stupendous chasm rivaling the Grand Canyon was carved out. Breeding ground and home today of great white sharks and many other forms of marine life, the Hudson Canyon was for a brief time a scenic wonder to the small, nomadic bands of Native Americans who periodically visited the Manhattan region of the glaciated terrain on hunting-and-gathering forays.

COMES THE FOREST PRIMEVAL

New York is a microcosm of the growth of cities and of the relationship between cities and the natural world. But the New York area was also a microcosm of the larger-scale picture of glaciation and postglaciation developments, a small part of the larger sequence of events being played out from the Atlantic Coast to the Rocky Mountains.

Each glacial advance, with its wall of morainal rocks and sediments at its front, was fringed by tundra, just as the land around arctic ice today is tundra. But farther to the south, below that tundra, lay forest, sometimes called the "boreal" (northern) forest, or just "spruce forest." On today's map, this northern forest lies largely in southern Canada, although it dips down into the United States in some northerly reaches such as the Adirondack Mountains of New York and parts of New England.

Below this boreal forest lies the familiar mixed hardwood forests of the northern parts of the United States; farther south, the forests contain still other mixes of species adapted to southerly reaches. With each glacial pulse, these bands of different sorts of forests migrated southward (trees can indeed migrate—not individuals, of course, but species migrate by sending out seeds). So as the climate grew colder, the ice advanced and the tundra always remained directly in front of it—and all the other belts of vegetation dropped down, too, farther and farther south.

Naturally, when the climate warmed and the ice retreated, these vegetation belts migrated back up north. Not to suggest that the process worked smoothly, like clockwork, with intact bands of spruce and of mixed hardwoods marching south, then back north, all species in lockstep. In fact, it was not at all like that, since different species disperse at different rates. But in general, it is true that at the height of the glaciation, the sort of mixed hardwood forests still visible in New York in places was displaced all the way down to the highlands of Mexico.

One interesting side effect is that the southward retreat—and northerly rebound—of bands of forest ecosystems (and, naturally, the animal life adapted to these systems) happened at a slow enough pace that most plant and animal species of the Pleistocene (the Ice Age) avoided extinction simply by moving and continuing to live in familiar habitat, albeit in places where their ancestors never could have survived. Most of North America's native plants, as well as the smaller animal species, were here throughout the Pleistocene—surviving major episodes of climatic change by simply tracking their habitats' southerly, then northerly, movements.

Only the truly large animals, most of them mammals, failed to survive. Among the Ice Age mammals of North America (certainly including New York) were mammoths and mastodons. These are commonly found as fossils in the New York City region, still unearthed virtually every spring when excavations start for new housing developments and shopping malls. North America up until about twelve thousand years ago looked like a cold-adapted version of the Serengeti Plain—that's how varied and numerous, and downright African-looking, our large mammals were.

But then calamity struck, and within a few thousand years they were all gone. And it was we humans who did them in. Humans arrived for the first time (at least, the first time in great numbers—recently uncovered evidence suggests that some humans were present in North America during the last glaciation) about 12,500 years ago. Archeological sites showing unmistakable signs of butchering using flint knives to hack flesh away from bones make it abundantly clear that the new North Americans were hunters, just as the people who had not long before that time invaded Europe, Asia, and Australia were hunters. And wherever humans showed up for the first time in great numbers, the large mammals on which they lived—animals who had never encountered humans before and so had not learned to fear human hunters—were easy targets for the spears and arrows of the early, very efficient hunters.

Retreating glaciers leave exposed bedrock, often ground and polished, as well as the sands and rocks of the moraines and other deposits. But they leave little or no actual soil—and there is precious little in which a plant may gain a foothold. You can see the whole process in progress at Glacier Bay in Alaska: the mouth of the bay is a full-blown northern forest; but as you go up the bay toward the still-retreating glacier,[5] the forest becomes sparser and more scraggly. Then only

5. The glacier there is a montane glacier, a river of ice filling (and further carving out) a valley between mountain peaks.

grasses and a few aspens or alders cling to the thin soil—until you reach the land on the banks just beyond the glacial front, where nothing at all in the way of large-scale plant life meets the eye.

That's more or less what happened in New York: just as you can travel from Hudson Bay to the Mississippi and find the spectrum of plant biomes sketched out here, you can also (figuratively!) stand in one place while the climate changes and see the various plant communities passing through. Sure enough, as time wore on and the glacier retreated ever farther north, eventually the forest began to take over from the tundra, so that by twelve thousand years ago a full-blown spruce-forest environment covered the New York region. Apparently, big game was already scarce, at least in these forests around New York, for there is little or no archeological evidence supporting the idea that humans were in New York back then, even though they had been there earlier when the region was strictly tundra.

Pine began to dominate over spruce by around seventy-five hundred years ago, known from the change in pollen types that preserve so well in pond and bog sediments dotting the landscape. By sixty-five hundred years ago, the deciduous forest had returned: the familiar mix of maples, oaks, birches (and many other trees, including, of course, conifers) known to us in the wild stands of almost entirely second-growth trees still around us today. These are present even in places in New York City, such as the hemlock grove in the New York Botanical Garden.

By then, Native Americans had returned in numbers, hunting and fishing, relying on the abundant marine and estuarine shellfish, to judge from the shell middens they left behind. Oysters were common, and the presence of the shells of some marine species of mollusks indicates that Native Americans at that time were moving about, perhaps seasonally, to harvest their shellfish resources.

Melting ice slowly brought the sea level up to approximately its present level sometime between forty-five hundred and six thousand years ago. So that is when the New York archipelago assumed its present, familiar shape. Manhattan, of course, is an island—surrounded by "rivers" (estu-

Figure 27. Extensive oyster middens in northern Manhattan left behind by Native Americans who utilized this important food resource from the nearby estuary. European immigrants, too, were fond of oysters and developed a large export trade with Europe. But pollution took its toll, and by the late nineteenth century most of the quarter million acres of oyster reefs in the harbor were lost. Courtesy of the Division of Anthropology, American Museum of Natural History.

aries all), with the Hudson on the west, the Harlem on the northeast, the East River on the east, and New York Harbor (Bay) at its southerly tip. But Manhattan is merely the most famous island here; Staten Island and Long Island head the list of other large islands in the area. But there are also Liberty, Ward's, Randall's (the latter two now artificially fused into one island), Riker's, Roosevelt, Oyster, Governor's, North and South Brother Islands, and many, many other smaller islands, some just small chunks of rock exposed at high tide. Not as prodigious an archipelago as Stockholm's, to be sure, but a bona fide archipelago nonetheless.

Thus the forest primeval. Not to suggest that Native Americans did not leave their mark on New York's landscape. Their footpaths and shell middens linger. And their hunting activities were a part of the larger picture that wreaked such havoc on the larger species of Ice Age mammals throughout North America. But the impact of Native Americans in terms of the out-and-out transformation of the physical landscape, introduction of alien species, and alteration of post–Ice Age, newly established ecosystems was nothing compared to the onslaught that began with the earliest settlers from Europe.

CHAPTER THREE

Landscape Transformed

Up to the point of sustained European contact, Manhattan Island and its surroundings offered a vista of rocky, hilly, and deeply forested terrain, with interspersed areas of wetlands and grasslands. Although Native Americans may have cleared land for settlements and agriculture here and there,[1] postglacial New York was as yet mostly unscathed by human hands. With the official advent of the Dutch settlement in 1624, all that was to change abruptly. What follows are vignettes of some of the most telling moments in the cutting, draining, and leveling that transformed the primeval forest and glades into settlements with agriculture and, ultimately, into the steel, glass, and concrete city we know today.

Little is known—and much disagreement persists—about exactly who "sold" Manhattan Island to the Dutch. Legend has it that the Native Americans, probably Munsees, living on Manhattan sold the land to Peter Minuit of the Dutch West India Company for trinkets (sometimes rendered as "twenty-four dollars' worth of junk jewelry"). Whatever the precise circumstances, the transfer was accomplished in

---

1. No direct evidence of precolonial Native American agricultural settlements has yet been found on Manhattan Island.

1626, a mere seventeen years after Henry Hudson first explored the region.

The lucrative fur-trapping industry (the colony's first official seal featured a beaver) spread like wildfire throughout North America and, by the early 1800s, had driven beaver and other fur-bearing mammals to the brink of extinction. Colonization of a few critical ports like New York immediately had far-reaching consequences throughout North America: the environmental impacts of major cities typically reach far beyond their obvious, immediate boundaries.

Naturally enough, the early European settlement, New Amsterdam, was concentrated on the southernmost tip of the island (an area that in due course would be expanded out into the bay itself as Battery Park and, much later, Battery Park City—see later in the chapter). Many of the homes in these settlements were the typical blowsy dwellings of the flotsam and jetsam of humanity who usually accompany the business-men setting up outposts far removed from home. And the lands of the poor reflected that—an instant squalor, harbinger of much of New York City's future, was established alongside the better-kept homes and gardens of the officers of the Dutch West India Company.

All this is simply to say that the early days of loosely planned and largely unregulated growth, of clearing land for farming, grazing, and firewood, of draining stagnant ponds and digging ditches, began to leave their mark on the island as soon as Europeans arrived in numbers. Their effects on the land were immediate, and the scars they left on the terrain went far deeper than the changes that the few, largely transient Native Americans who preceded them had wrought.

And so when we turn our attention to the details of the physical structural transformation of Manhattan and its outlying surrounds, we invariably find that the landscape to be leveled, drained, and otherwise reconfigured was often, if not always, already transformed into a bleak, and sometimes downright unhealthy, degraded ecosystem. Such appears to have been the case in the story of Manhattan Square, our microcosm within a microcosm.

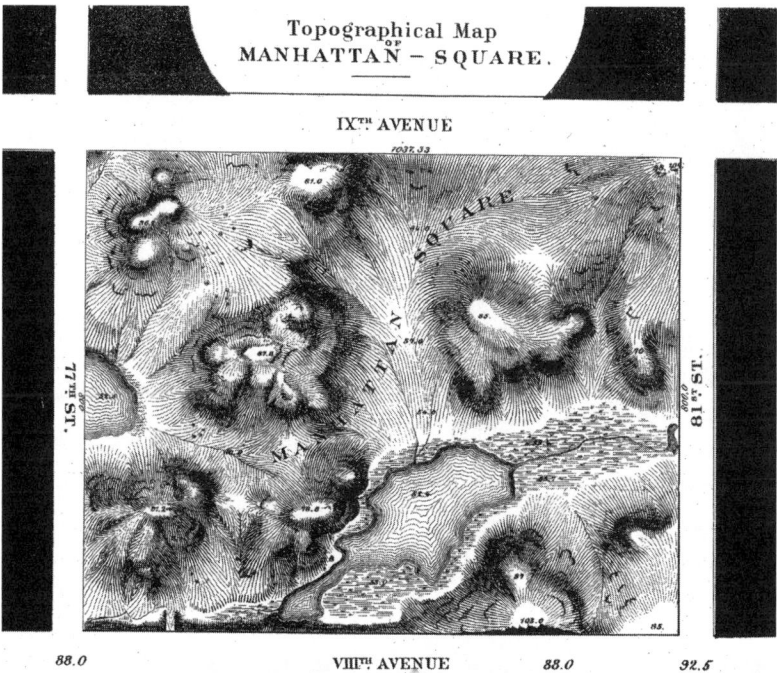

Figure 28. Manhattan Square, location of the future American Museum of Natural History. This map is a good portrayal of the site's irregular topography. Originally proposed as a site for a zoo, the rough topography was found unsuitable for that development. In preparing the site for the museum, builders had the hills cut down and the low areas filled with debris. Negative #325488, American Museum of Natural History Library.

MANHATTAN SQUARE

No place in New York City quite captures the essence of the upside/downside process of the construction/destruction of environmentally important institutions as well as Manhattan Square, a parkland measuring approximately seventeen acres and bounded by Central Park West and Columbus Avenue, and by West Seventy-Seventh and West Eight-First Streets. Known since 1958 as Roosevelt Park, Manhattan Square has been home to the American Museum of Natural History since the land was ceded to that fledgling institution by the commissioner of

Figure 29. The first building of the American Museum of Natural History, after the grounds were landscaped. This photo was taken several years after the museum opened in 1877. Negative #365466, American Museum of Natural History Library.

Figure 30. The American Museum of Natural History in 1877, before it was landscaped. Piles of debris from the surrounding demolished hills clutter the museum's grounds. Negative #00471, American Museum of Natural History Library.

Figure 31. The financial district, in southern Manhattan. With its curved, rather than parallel, streets the district retains some of its heritage from Dutch New Amsterdam and colonial New York. Photograph by Sidney Horenstein.

Central Park in 1872. The museum's first building opened to the public in December of 1877.

Manhattan Square was a by-product of an 1807 law mandating the laying out of most of the island in streets and avenues—the familiar rectangular grid of roads north of Canal Street that, for the most part, pays scant heed to the structural realities posed by Manhattan's underlying bedrock. Below Canal—where the city proper was still confined in the early 1800s—the roadways (including dear old Broadway) originally followed Native American trails-of-least-resistance around hills and fens, although a semblance of orderliness in the form of parallel streets soon appeared under Dutch, and later British, rule. But the streets of lower Manhattan looked nothing like the rigid grid work the Americans imposed above Canal Street.

Manhattan Square was one of the few chunks of land to escape the surveyors' thirst to impose a grid that disregarded obstacles in its projected path. The square was originally laid out within the boundaries of the grid, but because it was to be a park, one of several created on the grid plan, it was not subdivided by streets. Why? Simply because the square contained three prodigious hills and a pond. Hence Manhattan Square was designated nearly fifty years before Central Park was founded in 1857.

AROUND THE AMERICAN MUSEUM OF NATURAL HISTORY

One of the pleasant aspects of being associated with the American Museum is its setting. From a geological-ecological view, I think of the museum buildings as a granite outcrop surrounded by a tree-dotted grassy plain (somewhat savanna-like) within Theodore Roosevelt Park. As in similar settings not only in parks but also along streets, the trees represent both imported and native species. To complete this comparison, many of us view Central Park as the museum's backyard (and I suspect some New Yorkers do, too); countless numbers of "birders" have been led by museum ornithologists to explore its richness, especially during times of migration. One notable tree species in particular that stands out on the grassy lawn adjacent to the museum buildings is the very common London plane tree (*Platanus × acerifolia*), not because of the species' botanical features but because these particular specimens seem ready to topple over onto Central Park West. Just as in natural settings where trees are partly blocked from receiving a full range of light because they're growing behind rock outcrops or other trees, these hardy stalwarts lean away from the museum buildings, trying to get as much light as possible from the afternoon sun. Not to imply that this horticultural variety has no botanical features of interest, because surely it does, in addition to its historical connections.

Thus, it may be taken as a representative of many of the imported trees that populate our urban areas.

In South Lambeth, London, two gardeners, the Tradescants, father and son, were the proprietors of a famous nursery supplying plants to a wealthy clientele in England and the rest of Europe and were responsible for many of the trees and shrubs imported to North America. Upon his return from Virginia from a collecting trip in 1637, the son brought back specimens of the American sycamore (*Platanus occidentalis*) and planted them in his garden, which also contained the closely related oriental plane tree (*Platanus orientalis*), obtained at least by 1633 from somewhere in southeastern Europe or Asia Minor. So, two closely related species separated from one another by the expanse of the Atlantic Ocean grew alongside each other and hybridized as the London plane tree. Thousands of London plane trees with their telltale flaky bark were propagated from the Tradescants' small group of originals. As industrialization altered the environment across Europe during the nineteenth century, the hybrid proved to be especially tolerant of the coal dust, smoke, and compacted soil of cities. It was planted in numerous European cities, especially Rome, where it became a favorite tree of landscape architects who studied there.

In New York, the London plane was only sporadically used before the 1930s. Frederick Olmsted and Calvert Vaux, for example, do not even mention it in their Greensward Plan for Central Park. When Robert Moses created new parkways and planted trees in various venues in the 1930s, he hired landscape architects who admired the tree, and their enthusiasm caught on. Eventually these landscape architects established ninety thousand London plane trees in New York City, about 15 percent of the total number of trees and 30 percent of the canopy.

Across from the museum, another common street tree, the pin oak (*Quercus palustris*), is planted along the street on the park side of Central Park West. Constrained by the concrete roof and walls of the subway below, the trees have limited space to spread their roots, which rise up not only to wrap around the surrounding paving blocks and fold them into their roots but also to girdle the bases of trees, slowly causing their demise. And trees are not the only plants making an unusual and often difficult living in the city. Just south of the museum, the wall of Central Park, composed of somewhat porous sandstone from Nova Scotia, plays host to mosses and houses bacteria, algae, and surprisingly, ferns—especially where the wall is partly blocked from direct sunlight. Moreover, the crossbars on lampposts make nesting places for the imported house sparrows, and the museum grounds provide a feeding ground for starlings (*Sturnus vulgaris*), famously first introduced to the city in Central Park but now ranging across the entire country and Mexico and Canada, too. Appropriately, an early nesting pair of starlings was found on the facade of the museum in 1893.   *SH*

The American Museum of Natural History was initially the dream of its founder, Albert Bickmore, who had studied biology with the great Louis Agassiz at Harvard. New York had long wrestled with its bouts of poverty and lawlessness—the draft riots of 1863 were still fresh in memory. But despite the dangers and squalor, poverty and illness, cities have always been considered the acme of civilization—the highest state that human culture, exemplified in the arts and architecture, can hope to attain. Cities like New York are founded for economic gain, and New York, with its magnificent natural harbor and waterway leading to the interior, had been profitable from the very beginning, albeit to the relatively few individuals and companies who took the entrepreneurial lead.

But profitable it was. And that meant the emergence of a moneyed elite. (Mrs. Astor's list of four hundred socially prominent people springs to mind.) Yet you cannot have an aristocracy, even in a democracy, without the social and cultural accouterments that stamp a place—a city—as one of importance. Cities need cultural institutions to affirm their coming-of-age and bolster their collective self-image, as well as to educate and entertain their inhabitants (and later to help draw tourists; tourism currently ranks third among New York City's sources of income). San Francisco got its first of four nineteenth-century opera houses in 1864, not long after the gold rush brought a great infusion of inhabitants. Likewise an opera house was built in Manaus, an isolated city situated nine hundred miles up the Amazon, soon after the rubber boom put it on the map. (Construction of the Manaus Opera House began in 1884; it was not completed until 1899.)

New York City got *its* major opera house in 1880. Its first great art museum—the Metropolitan Museum of Art (the American Museum of Natural History's sister institution across Central Park)—was founded in 1870, a year after the natural history museum. The Philharmonic Symphony Orchestra was founded in 1878, the New York Public Library in 1895, the city's first major zoo (for years the Bronx Zoo, now more formally the Wildlife Conservation Society) in 1899, and the Bronx Botanical Garden (now the New York Botanical Garden) in 1891. Nor was this mere aping of hoity-toity European institutions (although Americans, of course, were historically Euro-centered, to the point of being Euro-wannabes; times have changed for many reasons): the major museums of Europe, for the most part, are only slightly older than their American counterparts.

And yes, part of the spectrum of desirable cultural institutions, then as now, is the natural history museum. At first simply the enlarged successor to the "cabinet of curiosities," the natural history museum before the age of movies, television, and even books with lavish color illustrations was the main connection between urbanites and the natural world that most of them had no hope of ever seeing "in the flesh." To this day, world-class natural history museums remain the best thing next to actually being in some real place in the natural world. Painstakingly preserved and mounted animals, displayed in dioramas built with pinpoint accuracy, are usually superior to the best that even modern zoos can manage.[2] Sadly, for some

2. Zoos, of course, have the advantage in displaying live animals; but seldom, if ever, are zoo denizens displayed in contexts that are 100 percent accurate—meaning that the plants and geological setting of an animal's native habitat are next to impossible to re-create, except in museum dioramas.

Figure 32. The Ninth Avenue elevated railway, about 1910. The original Ninth Avenue El opened in 1868, in lower Manhattan. When the American Museum of Natural History opened, in 1877, there was no station nearby. A plea from the museum's director worked, and the new station at Eighty-First Street opened in 1879. Negative #LS354013, American Museum of Natural History Library.

environments and species, natural history displays are often better than the degraded, even extinct, real thing: given the wholesale modification and out-and-out destruction of ecosystems the world over, virtually none of the habitats displayed in major natural history museums remain pristine and fully intact in the natural world.[3]

3. For example, of all the ecosystems around the world that are portrayed in dioramas at the American Museum of Natural History in New York, almost none

The impact that natural history museums have on the study of the natural world is immeasurable. Museums are truly libraries of biodiversity. The American Museum in New York currently holds over 35 million specimens and artifacts—of rocks and minerals recording the history of Earth itself; of fossils recording the history of life on the planet; of animals and microbes recording the present diversity of life on earth; and of human cultural artifacts, a reflection of human cultural diversity today and in our archeological past. In a kaleidoscopically changing world, where roughly thirty thousand plant and animal species are being driven to extinction each year, the collections of natural history museums provide the baseline for assessing the very state of the living world. The research and education missions of natural history museums grow more critical with each passing decade—a subject to which we return in a later chapter.

But it was the need to have exhibition space to serve the public that initially drove, and to this day largely sustains, the desire to have a natural history museum in the civic mind's eye. The gift of Manhattan Square to the American Museum's founding fathers was a mixed blessing: originally intended as the site for a zoo, the terrain, after becoming part of Central Park, was found to be far too rough to allow completion of the elaborate layout proposed as the zoo's design. Besides, plans for the American Museum were progressing well, and others thought the zoo belonged in Central Park in any case. Yet the landscape for the site of the prospective museum was bleak and daunting. Founder Albert Bickmore poignantly described the dismal situation:

> Within the boundaries of our area the prospect was most desolate and forbidding. There was a high hill at the northeast corner, of which a remnant still remains in the park[,] between Eighty-first and Eighty-second Streets;[4]

---

survive in the real world in the state originally photographed, painted, and carefully reconstructed in the museum's halls.

4. By "the park," Bickmore means Central Park. This outcrop still stands, a geological marvel whose top is sculpted by ancient glacial ice, while near its bottom a thrust fault reflecting tectonic events that took place hundreds of millions of years ago lies exposed to the initiated eye.

and in the north west corner [i.e., the present-day corner of Eighty-First Street and Columbus Avenue] another hill of solid rock rose much higher than the elevated railroad station, which now stands in its place. In the southern and central part of the square, just where the first section of our building was to be erected, was a third hill, whose crest rose as high as the ceiling of our present Hall of Birds. As I sat on the top of this rock, the surrounding view was dreary and my only companions were scores of goats.

There was, as well, a pond nestled among the hills, a pond that collected drainage from the north-south-trending ridge to the west of Manhattan Square (the ridge is still there today, as anyone walking up the incline from Columbus to Amsterdam Avenue is well aware). This drainage continued to flow as a small stream into the primordial form of Central Park's famous lake and eventually out to the East River. These were formidable obstacles to the construction of any building of size and dignity, let alone the behemoth proposed by the architects Jacob Wrey Mould and Calvert Vaux (the latter was the same man who, with his partner Frederick Law Olmsted, had so recently won the competition to design Central Park).

Vaux and Mould's design (figure 33) was never completed. Had it been, the entire original terrain of Manhattan Square would have been utterly transformed—much the way most of Manhattan Island itself has been completely transformed save for the parklands, which themselves are all designed and sculpted. The central tower was never built. But the southern facade, running along Seventy-Seventh Street between Central Park West (the former Eighth Avenue) and Columbus Avenue (Ninth Avenue) to the west, was built pretty much as originally conceived—and a version of the facade along Central Park West was also built. Today, the museum consists of a maze of interconnected buildings not quite so grandiose as Vaux's original scheme, but impressive enough. Parkland, especially on the northwest sector of what is now Roosevelt Park (including Margaret Mead Green), stands as a welcome respite to the unremitting exteriors of the buildings long since

Figure 33. The American Museum of Natural History, as originally planned. This layout was not completed, but today the complex is composed of twenty-seven interconnected buildings. Negative #3922, American Museum of Natural History Library.

Figure 34. Margaret Mead Green, situated in Theodore Roosevelt Park (formerly Manhattan Square), home of the American Museum of Natural History, between Seventy-Seventh and Eighty-First Streets, and between Columbus Avenue and Central Park West. New York's city council enacted a law in 1979 naming the northwest corner of the park after the distinguished anthropologist (1901–1978), who had been a curator at the museum from 1926 to 1978. Photograph by Sidney Horenstein.

erected in the Upper West Side—including the museum itself, to be sure, but also rows upon rows of apartments lining all the surrounding streets.

What a mammoth undertaking to get Manhattan Square in shape just for the original building (still aptly named "Building 1"), the building on Vaux's elevation (see figure 33) that was to have connected the proposed central tower with the southern facade. Bickmore recalled:

> As we wished to firmly locate our institution upon Manhattan Square at the very beginning and also to show from the first that our future edifice was ultimately to occupy its entire area, we decided to use our first appropriation in erecting one of the interior parts of the general scheme, and, therefore, we chose the wing radiating from the central tower or dome to the middle of the southern side as the first to be built, realizing that both of its walls must in time become only walls of interior courts.... The expense of cutting down the high hill of rock ...,[5] and of grading the surrounding area, proved to be so great, that we were obliged to defer using the selected red granite until we should begin to erect the exterior of the whole structure.

Hills and the pond were not the only obstacle to construction of the first building. Bickmore wrote that in 1872, "only the temporary shanties of squatters could be seen on the north, except two or three small and cheap houses halfway between Eighth and Ninth Avenues. On the west were only shanties perched on the rough rocks, and south of us there was no building near." Squatters would remain in and around Manhattan Square at least until 1893. The Upper West Side, with its rough terrain, goats, and squalid human habitations, was slow to be transformed.

---

5. Excavations in the inner courtyard of the museum in December 2003 briefly exposed the remnants of this third hill—beveled down just far enough to allow a smooth surface around the perimeter of the original building. The exterior of the southern and eastern facades was indeed eventually built of red granite, from a variety of sources. Old Building One is still very much in use. Its most famous exhibitions include the Hall of Northwest Coast Indians (first floor) and the Hall of Vertebrate Origins, opened in 1996, with much of its original architecture restored to its original state. The fifth floor of Building One houses the collections and offices of the Department of Mammalogy, one of the dozen research divisions of the present-day museum.

Figure 35. Strawberry Fields in Central Park, a tribute to the late John Lennon, who had lived in the nearby Dakota Apartment at Seventy-Second Street, where he was murdered on December 8, 1980. On March 26, 1981, the city council designated this area in Central Park as a memorial. John Lennon's favorite place in the park, and that of his wife, Yoko Ono, it is named after the Beatles song "Strawberry Fields Forever." Photograph by Sidney Horenstein.

According to Katherine Beneker, when the museum opened its first building in 1878, it "stood tall and bleak and unattractive amid the rocks and pools of the surrounding area."

The famous Dakota building—residence of the rich and famous, including the hapless John Lennon, who was murdered outside its gates in 1980—was built on the corner of Seventy-Second Street and Central Park West (then Eighth Avenue) and completed in 1884. It is said by some to have gotten its name because it seemed so far away from anything (that is, from everything that lay farther south in Manhattan) that it might as well have been located in the Dakota Territory. For a time, the museum's initial structure and the Dakota were among the few buildings looming over the hills and the piles of blasted rock that increasingly littered the landscape.

But spearheading the charge of northerly expansion, the Manhattan elevated railway, running along Ninth (now Columbus) Avenue, was opened in 1879. The neighborhood changed complexion quickly, although squatters' huts, undeveloped lots, and poor roads persisted even as late as 1887. Replacing this mangled terrain was the by-now landscaped Manhattan Square—with all hills now gone, and the lonely Building 1 lying midway between the avenues, close to Seventy-Seventh Street. And beyond Manhattan Square, to the north, lay buildings already lining parts of Eighty-First Street and seemingly all of Eighty-Second and other streets even farther to the north.

By 1892, the museum sported a facade on Seventy-Seventh Street; the building's footprint, at least, was faithful to the original Calvert Vaux design. By June 1900, the entire south facade along Seventy-Seventh Street had been completed. The east facade, along Central Park West, came later—including New York State's official memorial to Theodore Roosevelt—and was dedicated (by Franklin D. Roosevelt, then president of the United States) in 1936. The famed Hayden Planetarium (since replaced by the Rose Center for Earth and Space) had opened a year earlier, and it marked the museum's closest proximity to Eighty-First Street. The rest of Vaux's rectangular conception was never built, and while the museum has continued to add buildings, these have been placed within the existing architectural framework. The western and northern perimeters of Manhattan Square have remained as parkland.

Thus the yin and yang: build we must, and manifest destiny drove Manhattan Island's developers north. One of New York's greatest landmarks and most important cultural institutions—at first a dream, then an isolated and fragile outpost among the herds of goats and squatters' hovels—was connected to the already vibrant city to the south. Soon it would be joined by unbroken human habitation. We've gone from pristine beauty among imposing geological outcrops, to scenes resembling

Figure 36. The newest addition to the American Museum of Natural History, the Rose Center for Earth and Space. Composed of a glass box holding an eighty-seven-foot sphere that contains the Hayden Planetarium, the center opened in 2000, replacing an inadequate predecessor built in 1935. Within the structure is the academic and research facility of the Department of Astrophysics. © American Museum of Natural History/Dennis Finnin.

a war zone, to great architectural vistas. The transformation is stark, although the initial and final states each have their beauty. We lost a significant chunk of natural ecosystem, but Manhattan Square became home to one of the premier institutions on the planet, one that speaks out on behalf of the natural world.

COLLECT POND

The largest open body of water on Manhattan these days is the 106-acre, one-billion-gallon Central Park Reservoir, constructed between

1858 and 1862 in faithful conformance to Frederic Law Olmsted and Calvert Vaux's original Greensward Plan of 1858. South of that blatantly human-made, concrete-lined bowl of water lies the far more natural-looking "lake," while up near the park's northern boundary lies the Harlem Meer. Like much of Central Park, whose design incorporated many of the natural outcrops, hills, and depressions of the original landscape, the latter two bodies of water look deceptively as if they have always been there. But they haven't: they, too, owe their very existence to human design and clever engineering.

In truth, the title of "largest body of freshwater on Manhattan Island" must go to a long-since-drained-and-filled-in, sixty-foot-deep, now-forgotten lakelet originally dubbed Fresh Water Pond, known at the time of its demise in the early 1800s by the name of Collect Pond. This pond lay south of (what was to become) Canal Street, a block or so east of Broadway, with Chatham Street to the southwest. Today, Centre Street runs right through this former pond, and the infamous prison colloquially known as "the Tombs" occupies its western side. City hall was built nearby, and Foley Square (with its U.S. District Court and State Supreme Court buildings) was constructed on the site of the former pond. The southeastern part of the freshly drained and filled-in Collect Pond became the infamous Five Points neighborhood—where the draft riots took place and, more recently, the movie *Gangs of New York* was filmed.

Collect Pond, together with its diminutive relative Little Collect Pond, originally drained out to the East River and the Hudson River as well. (Thus the pond drained in two directions, but the East River connection was first to go.) The marshy Lispenard Meadows ran westward from the Collect. In 1785, two firemen recalled that they ice-skated, as boys, from the East River over the ponds and on through the meadows, all the way to the Hudson. Thus there's actually some debate about how fresh Fresh Water Pond/Collect Pond really was—a debate enhanced by the presence of large mounds of oyster shells left by earlier Native American inhabitants along the western side of the Collect.

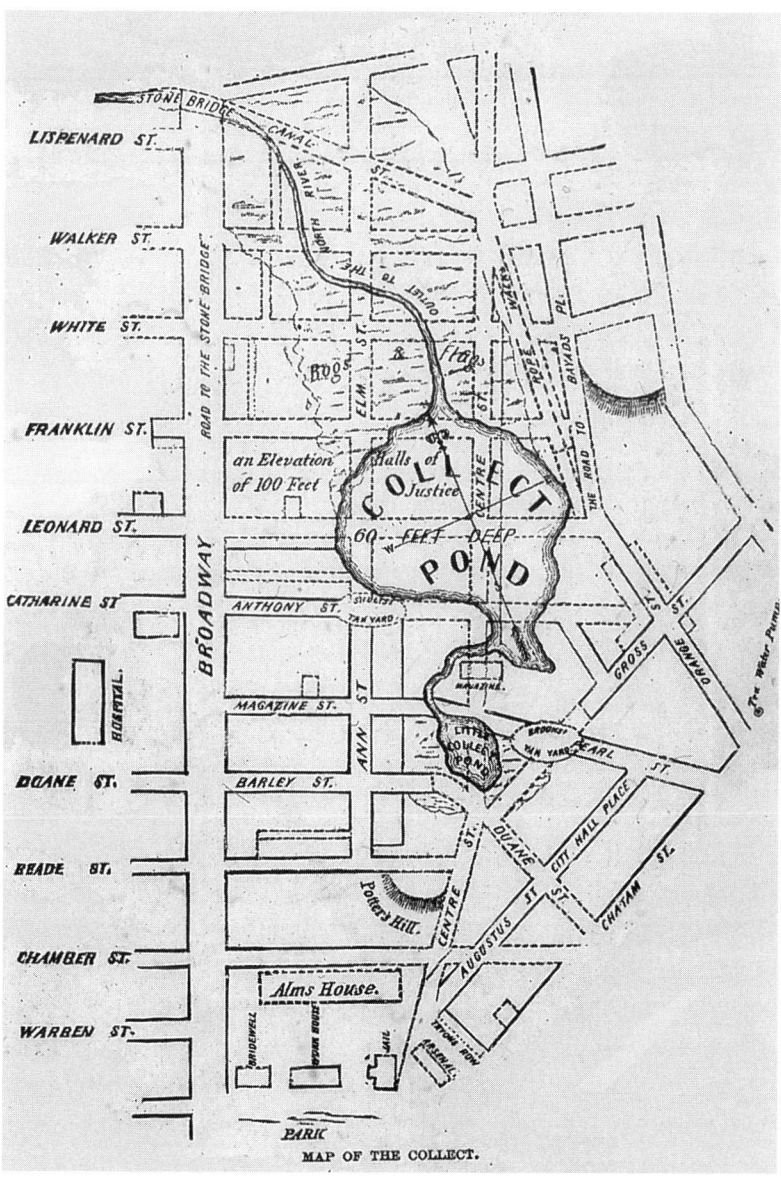

MAP OF THE COLLECT.

Figure 37. Collect Pond and Little Collect Pond. Collect Pond was a forty-eight-acre freshwater pond up to sixty feet deep that was nestled between former hills composed of glacial debris. The hills are now gone and so is the pond. Originally used by New Yorkers for picnics, ice skating, and freshwater, it eventually became overwhelmed by the growth of the city and was completely filled in by 1811. From *History of the City of New York,* by Martha Lamb (Barnes & Company: 1896).

Figure 38. Foley Square in lower Manhattan, site of the former Collect Pond. City, state, and federal courts, as well as associated prisons, are now located here. Photograph by Sidney Horenstein.

Oysters cannot survive in freshwater, but their presence suggests two possibilities: either the oysters were actually harvested from the pond, or they were brought there in prodigious numbers.

But the balance of the evidence favors the claim that the Collect was a thoroughly fresh body of water. There's the original name, of course. And as to the oyster shell middens, well, people went to the banks of the Collect on picnics in summertime throughout most of the eighteenth century, and the Native Americans most likely did the same with their oysters and other foodstuffs. But the most compelling reason to conclude that the pond was aptly named (as Freshwater; *Collect* evidently comes from the Dutch word *kolch*, meaning "a small body of water") is simply that, according to reports, people drank it—at least until it was no longer safe to do so.

In the mid-eighteenth century, Collect Pond lay at the north end of the as-yet undeveloped city. (Indeed, it is said that the south, east, and west facades of city hall, built between 1803 and 1812, were, as the sides that faced the denizens of the city, covered in resplendent marble; but the north side, facing the pond, where no one was, was instead covered with the much drabber, cheaper brownstone.) Evidence that the pond was then still very much a source of potable water comes in a round-about way from unexpected quarters. The Swedish Academy of Sciences (founded in 1739) adopted, as one of its goals, the discovery and importation of foreign plant seeds of potential use back home. To that end, they dispatched one Peter Kalm, at the suggestion of one of the greatest figures in the history of biology, Karl von Linné (Carolus Linnaeus, 1707–1778). Linnaeus published the tenth edition of his monumental *Systema Naturae* in 1758, the great work that established for all time Western science's approach to the study and classification of all living things, from microbes to fungi, plants, and animals.

Kalm had been Linnaeus's student, and Linnaeus eventually named the American mountain laurel (presumably Kalm's discovery insofar as his famed Swedish mentor was concerned) after him: *Kalmia latifolia.* In the November 2 entry in his 1748 diary, Kalm wrote of New York's waters: "There is no good water in the town itself, but a little distance away from there is a large spring which inhabitants use for their tea and other kitchen purposes. Those people who are less particular in the matter use the waters from wells in town, though it is very bad. This want of good water is hard on strangers' horses that come to the place, for they do not like to drink the well water."

Whatever the precise location of Peter Kalm's "spring," Louis Pope Gratacap, in *Geology of the City of New York,* cites an early reference to Tea Water Pump: "The first mention we have of the use of the spring water from the site of the Tea-Water Pump is found in the diary of a traveller in New York, in 1748." The diarist had found the "once-celebrated Tea-Water Pump, long covered up and disused, again in use, but unknown, in the liquor store ... [at] 126 Chatham Street," which was near the corner of Roosevelt Street. It is thought that the source of this pump was none other than nearby Collect Pond. A sort of resort, the

Tea Water Gardens, grew up around the pump. In any case, as early as the 1740s, "tea water men" began importing water in barrels from this pump north of the city. This was of course a harbinger of things to come: as New York grew, so did its thirst for clean water. That need led fairly quickly to an extensive system of reservoirs and waterworks that reached its tentacles as far north as the Albany region—and jumped as well from the Hudson to the neighboring Delaware River drainage, a topic we pursue in greater depth in chapter 5.

But a nice freshwater pond, useful for picnics, ice skating, and fishing, and as a source for water, was doomed given the tremendous pressure to expand the city northward. By the end of the eighteenth century, Collect Pond was heavily polluted, as tanneries, breweries, and slaughter-houses—all industries in constant need of freshwater—crowded the southern and eastern banks of the no-longer-pristine pond. Probably destined to be drained and filled in anyway, Collect Pond, so polluted that it was beyond worth saving, disappeared between 1803 and 1811. To help drain off the water and marshy grounds west of it, a canal was built along the outlines of the stream that ran through Lispenard Meadows— a canal that was later filled in to become Canal Street.

But the Collect was not so much drained as it was filled. Some of the first fill material came from the excavations of glacial material in the course of constructing city hall. Nearby hills, including Hangman's and Catiemutz Hills, shed some of their glacial deposits for the cause. And Bayard's Hill (previously called Bayard's Mount and, before that, Bunker's Hill after the Boston hill), at a hundred feet or so Manhattan's tallest, contributed its glacial crest. Soon the workmen had Collect Pond filled and leveled. By 1813 Collect Street—now Centre Street— was extended northward right over its center, appropriately enough. By the 1850s the tops of all the surrounding hills had long since been planed down—and street level was several meters higher than the old, quickly forgotten level of Collect Pond.

Middle-class buildings were constructed on the filled ground as soon as the drain-and-fill process was complete. But the ground was still damp, and the fill was sufficiently uncompacted and unstable that the buildings settled and tilted only a few years after construction. Basements regularly flooded. The demographics soon shifted, and the old Collect became a home to impoverished American blacks and immigrants and a haven for prostitution. It became the infamous Five Points.

HELL GATE AND THE NARROWS

New York is where New York is because it's a port that has supported commerce through worldwide shipping since the early days of Dutch settlement. Today, ships from around the world enter the port through "the Narrows"—the constricted bit of seaway between Staten Island and Brooklyn that, since 1964, has been spanned by the Verrazano-Narrows Bridge. And ships have always done so, at least if they were light enough so they did not scrape bottom. For the Narrows is an eroded channel through the glacial terminal moraine that runs between Brooklyn and Staten Island. Only considerable dredging of the channels on either side of the Narrows itself allows the passage of today's large ships. The extensive sandbars on the ocean side of the Narrows are especially shallow—and by the mid-nineteenth century had become an absolute deterrent to the ships increasingly dominating the trade routes.

Until dredging operations began in earnest in the early twentieth century, the main access to New York harbor lay through Long Island Sound. In the 1840s and 1850s, New York was the nation's leading port, and shipping and commerce its biggest business. The preferred route was through Long Island Sound to the East River, and then on down southward to South Street Seaport, which was then the main component of New York Harbor. The East River, a tidal strait, was less prone to freeze over than the Hudson on the west side; the east shore of Manhattan was on the island's leeward side, offering protection from the prevailing winds, and the shoreline was gentle on the east, in contrast

Figure 39. The Verrazano-Narrows Bridge crossing the Narrows between Brooklyn and Staten Island. When it opened in 1964, it was the world's longest suspension bridge. Because of the bridge's length and the height of its towers, the curvature of Earth had to be taken into account when it was constructed. At present, it has the eleventh-longest main span in the world. Photograph by Sidney Horenstein.

to the cliffs over the Hudson side. Long Island Sound itself, also on the east side, provided a stretch of about 120 miles of relatively protected waters for the transatlantic voyage.

The East River approach to the harbor cut the trip between the United States and Europe by fifty miles—which even in the days of steam shaved valuable hours off the journey. There was only one problem: Hell Gate, a treacherous run at the confluence of the East and Harlem Rivers—a section of water festooned with reefs (rocky outcrops that lie both above and below the tidal surfaces). Hell Gate extends from the point where the two channels of the East River meet just north of Roosevelt Island, to the point where the Harlem enters the East River—and the East River veers northwestward to connect with

EAST RIVER SHORELINE

The east and west shores of the East River possessed—and still possess—some interesting features. To the north on the Queens side, former large stone yards and their mills, with their piles of granite and marble slabs, once provided material for some of the famous buildings in the city. Today a massive power station occupies some of the land on which the former stone businesses were located, and just as these stone facilities usually received their materials by vessel, the power plant similarly receives some of its fuel by oil barge. The Ravenswood No. 3 Generating Station was built by Con Edison in 1963–1965 but, due to deregulation, has subsequently been owned by KeySpan, National Grid, and TransCanada. The power plant can generate approximately twenty-five hundred megawatts of power, which is about 20 percent of New York City's electricity demand. To the north is a small park named after Dr. Thomas C. Rainey, who for forty years attempted to get a bridge built here—one that was eventually built in a different location and called the Queensboro Bridge (now the Ed Koch Queensboro Bridge). This park is sited where he envisioned the bridge's pier would be put up. Beyond that is Socrates Sculpture Park, located on what was, until 1986, an abandoned riverside landfill and illegal dumpsite. A coalition of artists and community members transformed it into an open studio and exhibition space for artists and a neighborhood park for local residents.

On the Manhattan shore are Rockefeller University and the New York Hospital complex. Rockefeller University, situated between Sixty-Third and Sixty-Fourth Streets, is a world-renowned center for research and graduate education in the biomedical sciences and related fields, founded in 1901 by John D. Rockefeller. And just to the north is the equally well-known New York Hospital (recently renamed New York–Presbyterian Hospital/Weill Cornell Medical Center), framed by the East Side Highway. Eventually the highway's side-by-side lanes double up, running one above the other, because the depth of the water and the swift currents of the East River and Hell Gate allowed little landfill to be placed. Above and partly on the deck of the highway is Carl Schurz Park, home to the official residence of the mayor of the city of New York, Gracie Mansion. Carl Schurz (1829–1906) was a German revolutionary, American statesman, and reformer, and a Union Army general in the American Civil War. He was also an accomplished journalist, newspaper editor, and orator. In 1869 he became the first German-born American elected to the U.S. Senate (from Missouri) and later was a U.S. secretary of the interior. His wife, Margarethe Schurz, was instrumental in establishing the kindergarten system in the United States.

Views of Hell Gate from Carl Schurz Park are compelling: you can actually see the swift currents grab vessels as they make their way through the East River. It had to be the scene of dramatic events when shipping through this waterway was more important than it is today. At its north end, Roosevelt Island abruptly ceases, a result of the extension of the 125th Street Fault, and it is here that the river's east and west channels merge, forming the broad Hell Gate. The geology is at the same time simple and complex. Because of the way the rock formations are folded and distorted, the older Fordham gneiss makes its appearance where the younger overlying Inwood marble has eroded away. Roosevelt Island is underlain by the more resistant Fordham gneiss; and on the shore area of Queens the Ravenswood granodiorite gives it its elevation, while schist is found in Manhattan. The east and west channels of the East River are underlain by the weaker Inwood marble. It turns out that the merging of the channels south of Roosevelt Island, and consequently the wide reach of the river there, is caused by the Fordham gneiss submerging below the East River. Here, because New York's bedrock is generally tilted to the south, the strong Fordham gneiss dips deep below the marble and therefore is not available to make a high place in the terrain. From a vantage point on the Queensboro Bridge, looking south, you can see that, as the river approaches the other East River bridges, it turns to the east. Why and when this happened is still a mystery, especially since the Fordham gneiss separating the two marble valleys continues below the surface through the bulge of the Lower East Side.

There are many sites to see in a southerly direction from the Queensboro Bridge, one of which is the United Nations complex, the center of international politics, health, and science, which makes an imposing picture fronting the river. The bridge has gone through many repairs, alterations, and additions over the one hundred years of its existence; however, it fulfilled its builders' original intent admirably. Before the bridge opened, the borough of Queens was mainly semirural and, for all practical purposes, still an isolated part of the city, one that craved a direct connection to midtown Manhattan. With the coming of the bridge, Queens, now open for development, saw an influx of population, a rapid expansion of its trolley system, the introduction of elevated train service over the bridge, and an expansion of the subway system, all of which began converting it from an area of lush farms, estates, villages, and a small town (albeit one that had post–Civil War industrialization along its shoreline and Newtown Creek). When the bridge opened in 1909, the population was 275,000. In ten years it had doubled, and by 1930 over 1 million people lived there. The "garden of New York" was no more. *SH*

Long Island Sound. By today's street-layout reckoning, Hell Gate runs along the FDR Highway between Ninetieth and One Hundredth Streets.

Hell Gate was literally hell on ships. It was composed of roughly twelve islands and reefs, bearing names like Frying Pan, Pot, Bread and Cheese, Hen and Chicken, (Negro) Head, Heel Top, Flood, and Gridiron. But that wasn't all, for had they appeared in calmer waters those rocky hazards could have been charted and marked to reduce the threat of running aground. What made Hell Gate particularly hellish was

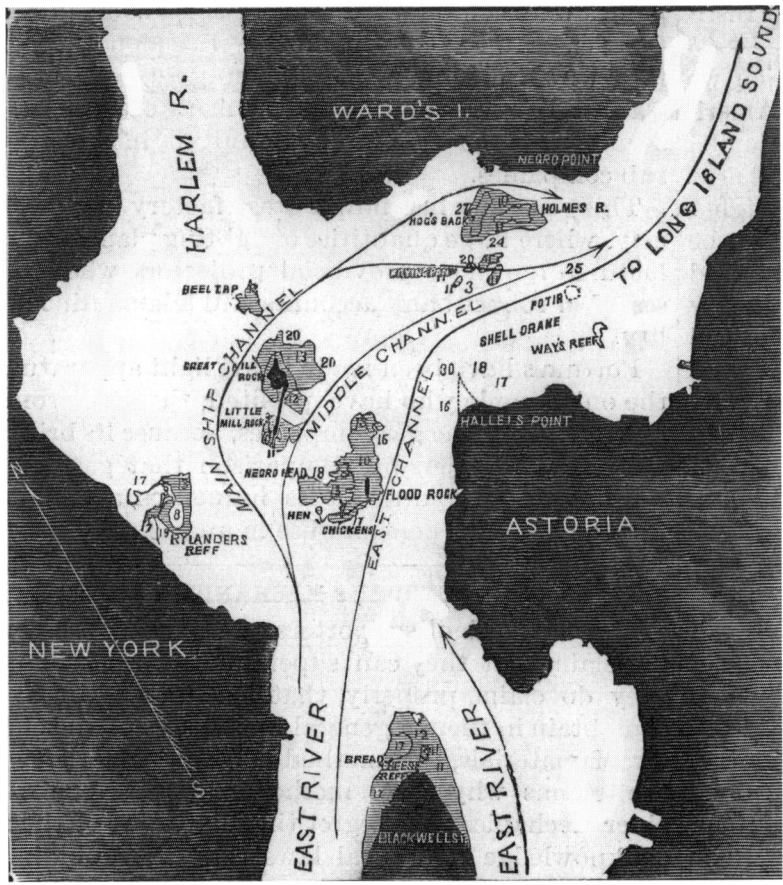

Figure 40. Hell Gate islands and shoals. This map was created before the U.S. Army Corp of Engineers removed the islands and shoals during the late 1800s and early 1900s. That effort eliminated navigation hazards and reduced the speed of the tidal flow in this formerly important commercial and recreational waterway. From "Blowing Up Flood Rock, Hell Gate," *Scientific American Supplement,* no. 522, 1886.

(still is, actually) its *tides:* it has two entirely different tidal regimes, one ebbing and flowing two hours before the other—one the Long Island Sound tide coming in through the connecting segment of the East River, the other coming from the ocean up the lower reaches of the East River. The Harlem River has its own tidal influences as well, and the nearly perpetual tidal motion, especially when exacerbated by winds, turned Hell Gate into a "dangerous cataract," in the words of one Captain Dermer all the way back in 1619.

During the Revolutionary War, Hell Gate was an obstacle to the British in the Battle of New York, forcing General William Howe to land his troops in Kips Bay south of today's Forty-Second Street rather than in Harlem. Worse, the British ship *Hussar* ran aground on Pot Rock—according to legend, it was laden with gold meant as payment for the soldiers but was sent instead to Davy Jones's locker. Salvagers are still desultorily looking for that gold.

By 1807, between five and six hundred ships sailed through Hell Gate weekly. The first commercial steamboat, Robert Fulton's *Claremont*,[6] successfully navigated Hell Gate's waters in 1815—a harbinger of things to come, as ships under their own power can maneuver in tight quarters much more readily than can large sailing vessels. Yet steam wasn't the entire answer, since many a self-powered ship hit the rocks as the nineteenth century wore on.

Periodically, merchants and other "concerned citizens" would raise an outcry over the dangers of Hell Gate. In 1832, a petition to build a canal through Hallet's Point ended instead with the New York State Legislature providing trained pilots to guide ships through Hell Gate. The petitions increased for the next fifteen years, and finally the federal government began a series of surveys that, by 1851, resulted in a detailed, accurate description of both the topography and hydrological conditions of Hell Gate.

But still the shipping woes continued. In the 1850s roughly one in every fifty ships ran aground—up to about a thousand ships a year. Something had to be done, and the campaign to clear Hell Gate began

---

6. *Claremont* is the name given to the boat sometime after it was first launched in 1807: in the beginning it was simply called "the steamboat." The first steamboat was not actually Fulton's but rather that of John Fitch, who tried out his new invention on the by-then heavily polluted Collect Pond in 1798—supposedly with Robert Fulton in watchful attendance. Another candidate for at least coinventor status of the steamboat is John Stevens, after whom the Stevens Institute, in Hoboken, New Jersey, is named.

Figure 41. Steaming across Collect Pond in 1798. John Fitch, who, like Robert Fulton, invented a steamboat, demonstrates the concept in Manhattan. After Fulton launched his commercial version in 1807, it revolutionized travel by river. From *History of the City of New York,* by Martha Lamb (Barnes & Company: 1896).

in earnest. The British had begun building large oceangoing steamships that could cross the Atlantic in two weeks, far faster than the fleetest of clipper ships. This alone threatened America's shipping interests, as approximately 70 percent of America's foreign trade in the early 1850s was still being carried by clipper ships.

To make matters more critical for New Yorkers, though, the larger ships that the British were now building *had* to go through Hell Gate to get to South Street Seaport. Back in the 1830s, when the maximum tonnage of ships was about a thousand tons, and the maximum draft did not exceed twenty feet, ships could wait "beyond the bar" outside New York Harbor for favorable winds and tides—a longer process than taking the Long Island Sound–East River–Hell Gate route, but a safer one. But by the 1850s, ships weighed as much as two thousand tons and drew as much as twenty-three feet of water. In 1855, the harbor commission wrote that "the depth of water on Sandy Hook Bar is 24 feet at mean low water, 23 feet at extreme low water. It is imprudent to send a ship of the largest class to sea except at high tide on the bar. The limit, then, for the passage of vessels at all states of the tide by way of Sandy Hook has been reached—while the tonnage of vessels and the draught of water is still increasing."

Hell Gate was the only option, yet the operators of the increasingly large ships were reluctant to risk their huge investments on negotiating this short but often lethal stretch of water. The earliest effort to physically alter Hell Gate was led by Benjamin Maillefert, who offered, for fifteen thousand dollars, to remove the most objectionable rocks with a new method of blasting described in his book. The chamber of commerce jumped at the offer. All went well as Maillefert and his crew relentlessly blasted away, leveling Pot Rock and greatly subduing an infamous whirlpool. But then disaster struck, when Maillefert was injured and several of his crew killed in a premature explosion in 1852.

The U.S. Congress stepped in, appropriating twenty thousand dollars to fund the Army Corps of Engineers (founded in 1775) to follow up on Maillefert's lead. But funds were soon exhausted, and Hell Gate kept on picking off ships. In 1856 an advisory council recommended removing the remaining obstacles, but nothing was accomplished before the Civil War broke out—and naturally nothing was done while the war raged on. Finally, in 1866, Lt. Colonel John Newton of the Army Corps of Engineers was assigned to examine the strait. Newton estimated that removing the obstacles would cost $3 million, which was not bad considering that the annual cost to shipping had risen to $2 million!

Initial attempts, though, failed, since the technique of underwater blasting from ships simply didn't work. Newton, by then a general, took over the project himself. To destroy Hallet's Point Reef, the general and his engineering corpsmen dug a deep pit below river level and then bored a series of tunnels into the reef. That process alone took seven years; by 1876, the walls and ceilings of the tunnels had seven thousand holes, four thousand of which were packed with a total of thirty thousand pounds of explosives. On September 24, with some onlookers predicting doomsday, the green shore of Manhattan filled up with people. Denizens of the insane asylum on Wards Island were led outdoors, while the prisoners on Blackwell's Island remained locked in their cells. There was a huge explosion, and the fragmented reef collapsed into the crater.

Immediately there was relief—turbulence decreased—and so, finally, did the shipping accidents. Newton turned then to Flood Rock and, with the same technique, got rid of that obstacle by 1885. By 1887, plans were in place to dredge Hell Gate down to a uniform depth of twenty-six feet. Hell Gate finally became safe for shipping—just in time to see modern dredging techniques begin to be applied to the shallow and treacherous offshore sandbars at the entrance to the harbor. By then the size and number of ships had increased so much that the ports were expanding throughout New York Harbor. The East River, Hell Gate or not, just couldn't take all the traffic!

BATTERY PARK

Not all the work of building Manhattan entailed leveling hills and filling in ponds and marshlands or blasting watercourses: Manhattan today is actually larger than it was when first settled—the direct result of filling in and building up the shoreline with chunks of bedrock and other debris. The most extensive of all these early projects was the creation of Battery Park and, much later, its extension, Battery Park City. Almost all of Battery Park, including Castle Clinton, lies on human-made land.

Manhattan's southern tip was already a popular place by the late eighteenth century; people took the fresh sea breezes and admired the view along a strip of land that had been created by adding the demolished remains of Fort Amsterdam (later renamed Fort George by the British), and the remains of the small rise on which it was built, to the shoreline. Before Fort George was demolished, people had strolled on

Figure 42. Castle Clinton in Battery Park, at the southern tip of Manhattan, originally built in 1808 as a fortification three hundred feet offshore. In the nineteenth century, after its usefulness as a fort ended, the structure was subsequently an entertainment center, immigration depot, and the New York Aquarium. Restored, it now serves as a reminder of past harbor defenses and as a ticket booth for ferries to Liberty Island. Photograph by Sidney Horenstein.

the gun emplacement wall (the Battery). This fort was constructed on the site of the what is now the Customs House.[7]

Named for the gun batteries emplaced on the southern edge of Manhattan in 1683, Battery Park began to grow at the expense of the harbor's waters in a series of episodes. First, in 1807, with the threat of renewed war with Britain, work was begun on Castle Clinton (named after

7. The Alexander Hamilton U.S. Customs House is currently owned and maintained by the Smithsonian Institution, and since 1989 it has been the New York branch of the Museum of the American Indian. The latter was formerly housed off Audubon Terrace on Manhattan's northwest side, just south of the George Washington Bridge.

Governor DeWitt Clinton in 1815), one of a series of fortifications that were to stand as prototypes of U.S. coastal fortifications for at least the next fifty years.

Castle Clinton was actually perched on a group of rocks that lay some three hundred feet off Capske Hook (a Dutch name meaning "Rocky Ledge") at Manhattan's southern tip. Construction began with dumping blocks of Late Triassic–age brownstone from New Jersey until the pile stabilized and the added rocks sank no farther around the periphery of the preexisting rocks. The walls were then built up, but they soon began to tilt, and work stopped after completion of the first tier, presumably because of the tilting, unlike at the four-tiered companion fort, Castle Williams on Governor's Island.

Since the fort was hundreds of feet offshore, a bridge was soon constructed to connect it to Manhattan—a bridge replete with the traditional draw near the castle's entrance. Castle Clinton was not actually used in the War of 1812,[8] but its bridge became a haven for fishermen. By 1823, the U.S. government gave the castle and its surrounds to the City of New York—which promptly turned it into an entertainment center, as the views and refreshing breezes continued to lure Manhattan's inhabitants out from the less agreeable parts of the city. Renamed Castle Garden in 1824, the building became an important New York theater with the addition of a roof in 1845.

Meanwhile efforts to join the fort to the city by terra firma continued apace, and by 1853 landfill had finally filled in the last remaining gaps between the two. In an abrupt change, the city then leased the building to the state. In 1856 Castle Clinton became an immigration center, processing around 8 million people before it closed in 1890. The federal government then took over and processed immigrants at the Barge Office in Battery Park, until the famous center on Ellis Island opened in 1892.

---

8. Apparently New York's fortifications were a sufficient deterrent, and the British decided to attack Washington, D.C., instead.

But life was still not over for the multiuse Castle Clinton. After the fort was placed once again in the hands of the city, the New York Aquarium opened on the site, in 1896, where it existed until 1941 as a much-beloved precursor to today's Coney Island Aquarium. At that time, Robert Moses (1888–1981), the famous, irascible planning and development czar who oversaw much of the city's later growth and development, earmarked it for demolition. Determined to use the space to build the Manhattan tower of the Brooklyn-Battery Bridge,[9] Moses began demolishing the castle. World War II co-opted his heavy machinery, interrupting the demolition, and none other than Eleanor Roosevelt joined civic groups in the charge against Moses's plans. After reading her article, the secretary of the navy declared that Moses's bridge would block the entrance to the East River, and thus to the Brooklyn Navy Yard. Apparently in a revengeful pique, Moses closed Battery Park to the public. Among other things, this made it difficult for the public to ascertain just how much of the fort had already been destroyed. As it turned out, the aquarium was destroyed, but the fort itself was not. Eventually, in 1946, Congress declared the fort a national monument—though Battery Park remained closed until 1952, while the Brooklyn-Battery Tunnel was under construction.

A fortification, an entertainment center, a theater, an immigration processing center, and then an icon of biodiversity, New York's much beloved aquarium: quite a potpourri of roles for this added-on outpost at Manhattan's southernmost tip. That Castle Clinton today serves as the ticket booth for the Statue of Liberty and Ellis Island ferries seems at first a humble position for such a noble structure to assume. But there it is, still commanding a magnificent harbor view, still with those refreshing un-city-like sea breezes—and serving as the jumping off point for trips to two of America's most iconic symbols of our global connections.

9. Moses disliked tunnels and said he thought of them as "vehicular bathrooms." To him, tunnels were holes in the ground, and bridges were monumental works to be admired. He opposed the structure that eventually was built—the Brooklyn-Battery Tunnel—in favor of his plan for a bridge.

Figure 43. The Statue of Liberty on Liberty Island, Upper Bay, viewed from the Staten Island Ferry. The statue was erected on the parade ground of Fort Wood, one of the coastal defense fortifications in the harbor (completed in 1811 and later modified). From the ground to the torch, the statue measures 305 feet 1 inch tall. It was dedicated on October 28, 1886. Designed by Frédéric-Auguste Bartholdi (the interior supporting structure was created by Alexandre-Gustave Eiffel), it was a gift to the United States from the people of France. Photograph by Sidney Horenstein.

And what of Battery Park City, built as a northward extension of Battery Park in the 1980s? The landfill beneath it came from the giant excavations for the foundations of the twin towers of the World Trade Center, completed in 1973 and demolished, as everyone sadly remembers, on September 11, 2001.

THE HARLEM RIVER SHIP CANAL

Unlike Manhattan's artificially increased southernmost tip, its northernmost tip is artificially diminished, another instance of the monumental transformation of landscape in the service of rendering New York and its environs more amenable to the way we human beings see fit to live. The construction of the shipping canal between the Hudson and Harlem Rivers literally separated a chunk of Manhattan—Marble

Hill—from the rest. For a time, the hill was completely surrounded by water: the ship canal lay along its south edge, and Spuyten Creek curved around the rest of the hill, which remained connected by bridges to Manhattan Island and the Bronx. Eventually Spuyten Duyvil Creek was filled in with construction debris from elsewhere, forming a land connection between the hill and the Bronx. Today, even though it is no longer attached to Manhattan Island, Marble Hill is still a part of Manhattan administratively.

Crossing the Henry Hudson Bridge these days, which passes over the Harlem River near its junction with the Hudson River, one sees little that suggests the vision is not natural but rather the outcome of a phenomenal engineering and construction project. One sees Columbia University's Baker Field and surrounding fields and buildings, home to Columbia's often struggling football program and other sports.[10] Across the river, in what is the Bronx, there is a large rock face with a big blue *C* painted on it—for Columbia, an acknowledgment that its rowing teams practice and compete in these often rough waters of the Spuyten Duyvil.[11]

That watery divide between the rock sporting the big *C* and Baker Field on Manhattan's now northernmost tip, and which everyone erroneously still calls Spuyten Duyvil, is no such thing. The Big *C* rock is actually the face of the north-trending ridge of Fordham gneiss that extends through the Bronx communities of Spuyten Duyvil and Riverdale to the Yonkers border. The real, original, S-shaped Spuyten Duyvil Creek once lay farther to the north and to the south of where that rock stands; the

10. We make this statement with some heartfelt angst; one of us (NE) was a trumpet player in the Columbia College band the last time Columbia won (shared, actually) the Ivy League title. That was in 1961!

11. *Spuyten Duyvil* is Dutch for "in spite of the devil." According to Washington Irving's story, the New Amsterdam garrison trumpeter Anthony Van Corlaer, who was charged with warning the Hudson River villages north of the island of the British invasion—with the help of his horn and a ferry—found the northern ferry point unmanned. After a deep swig of "Dutch courage," he attempted to swim across "in spite of the devil." Legend tells us that he was caught by Old Nick and drowned midstream, but not before his last warning note was sounded.

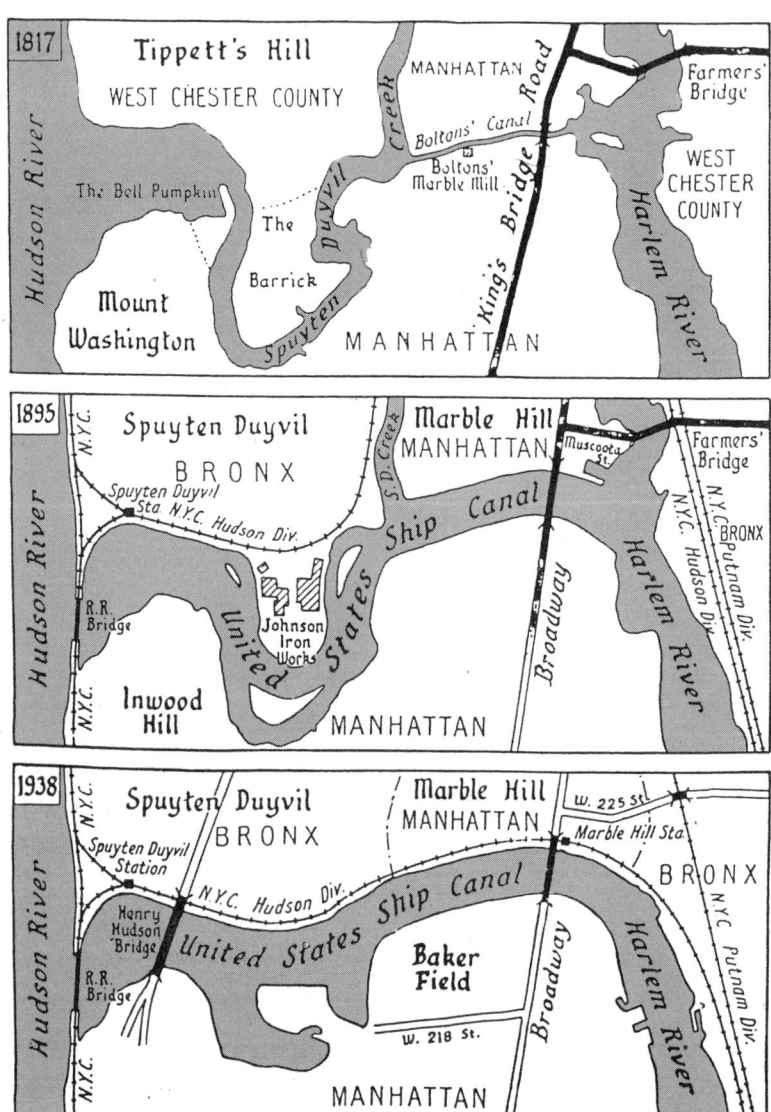

Figure 44. The alteration of the Harlem River and Spuyten Duyvil Creek to create the U.S. Ship Canal. The first section of the project, completed in 1895, cut off the north loop of Spuyten Duyvil Creek and deepened the waterway to enhance commercial marine activity. The project was completed in 1938, when the southern loop of Spuyten Duyvil Creek was eliminated. From the Bronx County Historical Society Collections.

Figure 45. Looking north to the Spuyten Duyvil section of the Bronx. The cliff with the graffito *C* (for Columbia University) was created by the excavation of the final phase of the U.S. Ship Canal in 1938. Photograph by Sidney Horenstein.

ship canal was cut through the middle of the S, which is where the rock with the painted *C* stands. The Hudson Division of the Metro-North Railroad is routed on the north side of the ship canal, and when it reaches near the Hudson River it is joined by the Amtrak line. (See figure 44.)

Though the Spuyten Duyvil did form a watery connection between the Hudson and Harlem (making Manhattan an island ever since the last glacial retreat allowed the sea level to rise sufficiently to flood the channels), it was far worse a place than Hell Gate as far as shipping was concerned. Indeed, you could wade across when the tide was low. To make matters worse, the waterway was S shaped, and it seemed almost to cry out to be straightened as well as deepened. After the Erie Canal—a trade route to the interior—was opened in 1825, Governor DeWitt Clinton wanted to make a direct connection between it and New York's South Street Seaport. But although the Harlem Canal

Figure 46. The swing railroad bridge over the Harlem River at its juncture with the Hudson River. This bridge replaced the original wooden structure that was opened in 1849 as part of the New York and Hudson River Railroad. Today the bridge forms part of Amtrak's Empire Corridor line between New York City and Albany. Photograph by Sidney Horenstein.

Company was founded in 1826, no progress was made. Nor did anything much come of desultory efforts throughout the 1860s.

But in 1873, Congress ordered a survey with the idea of a canal in mind. New York State purchased the land and turned it over to the federal government (the feds had said they would not give restitution to the existing private landowners because their land was condemned). Various proposals were floated, including an especially ambitious one by a Mr. Charles Stroughton calling for a 15-foot-deep, 350-foot-wide channel. The route would extend to Europe—if, that is, Port Morris in the Bronx were to become the new New York Harbor (not a bad idea given the ongoing perils of Hell Gate).

Work finally began in January 1888 on what was formally known as the United States Ship Canal, but otherwise known locally as the Harlem River Ship Canal. The plan by then had expanded, and a channel 400 feet wide and 15–18 feet deep would be cut through rock at Dyckman's Meadows to provide a straight route for Spuyten Duyvil Creek. The first cut through Marble Hill was made in 1895, and the other cut—which isolated the rock that today has Columbia's *C* festooning its side—came in 1938. The old Spuyten Duyvil bed was eventually filled in with train-delivered rocks removed for the foundation of Grand Central Station—all the way down on 42nd Street in central Manhattan.

None of this came easy. Protesters against the U.S. Ship Canal proposed alternatives, such as filling in the southern part of the Harlem River—thus shelving the idea of shipping in favor of connecting Manhattan with the mainland to the north. There is little to suggest that commerce was greatly enhanced by the shipping canal project, beyond making Manhattan a thoroughly circumnavigable island that operators of the Circle Line and other tourist lines can exploit as they ferry the hordes of people who have made New York a favorite destination.

These vignettes of the monumental alteration of the original landscape in building New York convey the general flavor of engineering projects that have transformed a modestly hilly terrain—with woodlands, streams, and ponds scattered about—into what all but the most discerning pedestrians tend to think of as a relentlessly flat surface. Murray Hill, though, still persists, as does Harlem's Marcus Garvey Park. Manhattan was not rendered utterly featureless as the relentless northward expansion demanded flat surfaces for its regular grid of roadways and for the foundations of its buildings.

And the picture that emerges from these larger-scale, more heroic attempts to tame and smooth the landscape for human use reveal a complex mélange of serious plan and accidental happenstance, of economic necessity and a sort of serendipity, that seems to be at the heart of the complicated reasons why people do things in general. Hell Gate had to go in order for New York shipping to survive—but once the problem had been solved, underwater dredging technology appeared, and the way through the Narrows was suddenly possible. New York needed a fort to ward off British attack, but the fort never came under fire in the War of 1812. And the fort building, still standing, went through an amazing sequence of uses that seems, in retrospect, far more adventitious than the result of careful advanced planning. Collect Pond, once a source of water and a haven for ice skaters and picnickers, became polluted and a literal impediment to further northward expansion of the city, so that the only solution really was to get rid of it. But it was done improperly, and the leaning, cracking buildings soon became slums—and literally sank, like the low status of the infamous Five Points section. Now the site is home to Foley Square, and there the New York Supreme Court and the New York Federal District Court stand out among many other bastions of local, state, and federal governmental authority. And finally, the story with which we began this chapter—the transformation of Manhattan Square into the locus of one of the world's major natural history museums—also blends the quirkiness of chance with the determination of dedicated people.

And so the way is paved for us to chart the exponentially accelerating pace of growth—of buildings, of institutions, of people—that has brought New York City to its present preeminent place among the cities of the world.

CHAPTER FOUR

Growth of the Concrete Jungle

When Beatle John Lennon was murdered on December 8, 1980, the news got out fast. Fans flocked to the scene of the shooting—the Dakota, that enormous apartment complex where Lennon lived with Yoko Ono and their son, Sean, along with other celebrities and wealthy Manhattanites. Now a standard stop on tourist bus routes, the Dakota still looks a bit more impressive than its surroundings. But the building, standing at the corner of Seventy-Second Street and Central Park West—down five blocks from the even more massive American Museum of Natural History and across the street from the world's most famous park—is far from being the isolated early outpost of grandiose living it was when it opened in 1884. When Lennon died, the world instantly heard about it simply because the Dakota, while still affording a measure of opulent privacy within its walls, now lies near the very heart of the most dynamic, media-saturated city in the world.

People still argue over the origin of the Dakota's name. Some pronounce it a myth that the Dakota is so-called because its isolation on the little-developed Upper West Side when it was first built—a century before the violent death of arguably its most famous tenant—brought to mind the isolation of the Dakota Territory far to the west in a still-expanding America. Some prefer to think that New Yorkers,

Figure 47. The Dakota Apartment, Seventy-Second Street and Central Park West. Completed in 1884, when development began in this part of the city, the building is a National Historic Landmark as well as a New York City Landmark. Photograph by Sidney Horenstein.

already growing tired of the dreary paving stones, bricks, and building stones that dominated their daily lives were already yearning for the bucolic vistas of untamed lands and simply developed a preference for evocative names of far-off places. (Though this may have been only in the mind of the Dakota's builder, Edward Clark, who, more than six months before the plans for the Dakota were filed, had also proposed renaming Ninth and Tenth Avenues as "Wyoming Place" and "Arizona Place," respectively.) After all, Central Park was developed with just this spirit in mind. And the soft yellowish pastel color of the building stones of the Dakota's facade (300-million-year-old sandstone brought at great expense from Nova Scotia, when plenty of other types of building stones lay closer to hand) does seem to impart a sense of the great outdoors, as some architectural critics and historians have claimed.

Clark was a developer who spent $5 million buying up former farm-land on the Upper West Side in the 1870s. Though he built a tract of row houses along Seventy-Third Street, as well as tenements along Ninth Avenue (Columbus Avenue since 1890), there's no question that the Dakota was his pièce de résistance. The idea of housing a number of wealthy households under a single roof apparently came from France, having first arrived in New York in the form of the Stuyvesant Apartments, completed in 1870 by Richard Morris Hunt.

However daring and original was "Clark's folly" in providing a common dwelling for the rich, one that was not too far from middle-class and working-class housing in the largely undeveloped Upper West Side, such speculative ventures lay at the heart of the effort to stake out new territories as the expansion of Manhattan continued apace in the latter half of the nineteenth century. Clark's project was a microcosm of the entrepreneurial spirit that saw to the completion of the concrete jungle.

NEW YORK'S CITY PLAN

As nearly every new arrival to Manhattan soon finds out, the maze of intersecting streets in Greenwich Village, Soho, Tribeca, the financial district, the Lower East Side, and Chinatown, at first so difficult to negotiate, gives way to a far more orderly grid system of numbered streets and avenues above Houston Street (or above West Fourth or, finally, above Fourteenth Street—depending which side of the island you are on and how regular you need your grid to be). Actually, the map of lower Manhattan shows not the chaos that a neophyte feels upon first navigating the streets but a series of fairly regular gridlike sectors, but ones that nonetheless come together at odd angles and give the feeling of disorder. Relatively few of Manhattan's streets are sinuous, although its most famous, Broadway, which somehow managed to survive after the northern grid was superposed on the terrain, is still charmingly bent in places as it wends its unique way along all of the

Figure 48. An example of the irregular street pattern in northern Manhattan. The irregularity is the result of a diverse topography that did not fit the grid system, one of the reasons the original grid system ended at 155th Street. Photograph by Sidney Horenstein.

island. Curved streets such as Riverside Drive and Edgecombe Avenue were, however, created after the grid was imposed, especially along the east and west sides of the island.

New York was by no means the first American city to adopt a gridiron layout for its streets and avenues. By the end of the eighteenth century, Baltimore, Cincinnati, Savannah, New Orleans, and Philadelphia had all gone this route. Order and simplicity seemed to be the main virtues of such plans, but difficulties with terrain and, not least, preexisting patterns of land usage and ownership have provoked dissent virtually everywhere grids have been adopted.

Philadelphia's plan was perhaps the most influential of all, because immigrants passing through that city, before New York became the premier port of call for immigrants, were struck by the simple street plan and helped disseminate such layouts in cities and towns across growing

America. It's not clear, however, where the idea to lay New York out in a similar gridlike fashion came from. What is known is that land on Manhattan owned by loyalists during the Revolutionary War was confiscated under various acts, which were passed on various dates, and this resulted in the abandonment of New York by forty thousand Tories between 1782 and 1783. In 1785, these common lands, together with privately held property—in roughly the area of what is now Manhattan south of Canal Street—were surveyed by one Casimir Goerck (the city surveyor), who mapped them in a rectangular pattern. A second survey followed in 1796, encompassing a wide strip through central Manhattan, with north-south streets a hundred feet wide, and with cross streets sixty feet wide and spaced at regular two-hundred-foot intervals. All grids ever proposed for Manhattan always had, of course, far more cross streets than north-south thoroughfares, not only for the obvious reason that Manhattan is a good deal longer than it is wide, but also, and more subtly, to yield ready and consistent access to the Hudson and East River shorelines, the focus of commerce in those early days.

The problem bedeviling the maps of this second survey, though, was that they did not closely agree with property descriptions given in privately held deeds. To keep local politics and conflicting claims of property owners from snarling the process, the city asked the state legislature to appoint a commission—the Commissioners of Streets and Roads in the City of New York—and suggested the names of Simeon DeWitt, Gouvernor Morris, and John Rutherford. On April 3, 1807, the measure was passed, and the three commissioners were approved with the following duties: "to lay out ... the leading streets and great avenues, of a width not less than 60 feet, and in general to lay out said streets, roads and public squares of such ample width as they may deem sufficient to secure a free and abundant circulation of air among the said streets and public squares when the same shall be built on." The map was to be "final and conclusive."

The commissioners also found it impossible to adjust their plan to the irregular property boundaries and the random streets that already existed. In the Commissioners' Plan of 1811, the report that was made public in 1811 (including the map, authored by John Randel Jr.), they stated the reasons for adopting the gridiron plan:

> That one of the first objects which claimed their attention, was the form and manner in which the business should be conducted; that is to say, whether they should confine themselves to rectilinear and rectangular streets, or whether they should adopt some of those supposed improvements, by circles, ovals and stars, which certainly embellish a plan, whatever may be their effects as to convenience and utility. In considering that subject, they could not but bear in mind that a city is to be composed principally of habitations of men, and that straight sided, and right angled houses are the most cheap to build, and the most convenient to live in. The effect of these plain and simple reflections was decisive.

Thus, whatever the influence that Philadelphia and other gridiron-roadway cities may have had on New York, the famous radiating plan of Washington, D.C., was explicitly and decisively rejected!

The 1811 plan called for 155 streets at least sixty feet wide, which would run at a right angle to a dozen north-south-oriented avenues. And though they obviously based their plan on the previous Goerck plan, they did not acknowledge the latter in their report. People were surprised that the plan extended so far north of what was then the developed city—all the way up to 155th Street, and with an extension of Tenth Avenue to the northern end of the island. In their report, the commissioners wrote, "To some it may be a matter of surprise, that the whole island has not been laid out as a City; to others, it may be subject of merriment that the Commissioners have provided space for a greater population than is collected at any spot on this side of China. They have in this respect been governed by the shape of the ground.... To have come short of the extent laid out, might therefore have defeated just expectation and to go further, might have furnished materials to the pernicious spirit of speculation." At 155th Street, where the plan stopped, the topography changes dramatically. However, the commissioners did not let abrupt elevation changes bother them as they mapped farther south, in areas such as the ridge (Morningside Heights) adjacent to the Harlem Flats.

Figure 49. Bennett Park, at 184th Street and Fort Washington Avenue. The top of the rock outcrop situated here is Manhattan's highest elevation, 265.05 feet. The park was the location of Fort Washington, captured by British and Hessian troops on November 16, 1776, and occupied by them until the end of the Revolutionary War, in 1783. Photograph by Sidney Horenstein.

The commissioners were met with hostility by property owners. Surveyor John Randel Jr. wrote about his experiences decades later, in 1864, saying that "Col. Richard Varick [a former mayor]...always became bail for my appearances at court, when, in the absence of the Commissioners, I was arrested by the Sheriff, in numerous suits instituted against me...for trespass and damage committed by my workmen, in passing over grounds, cutting off branches of trees, &c." He remembers being pelted by artichokes and cabbages during one encounter; a law was eventually passed to protect him from further suits. The appointment of the commission of course led to land speculation; and while some owners objected initially, benefits accrued to families, "many of whose descendents have been made rich thereby."

ONE HUNDRED AND FIFTY-FIFTH STREET

The Commissioners' Plan of 1811 laid out the city streets up to 155th Street, with an extension northward along Tenth Avenue (Amsterdam Avenue) to the end of the island. One day I set out to explore 155th Street to gain insight into why the plan stopped there. Entering the street from the Eighth Avenue subway station at 155th Street, I was surprised to see how wide the street was that led eastward downslope to the long viaduct bringing pedestrians and vehicles to the Macombs Dam Bridge over the Harlem River. The viaduct itself was needed because of the substantial drop-off in the terrain toward the Harlem Valley and the bridge. Predecessors of the current bridge have a long history, first making an appearance in 1839 as a dam and bridge, which caused a lake to form in the valley extending to the north. Here is the beginning of Highbridge Park, where streets do not follow the grid pattern. Edgecombe Avenue rises as it curves around the park, coming to Coogan's Bluff and the historic Morris-Jumel Mansion; to the east a road sweeps downward in a broad curve to Harlem River Drive.

As I walked westward along 155th Street toward the Hudson River, I was impressed by the narrowness of Manhattan Island here compared to the much greater width farther south. One of the purposes of my visit was to see the location of John James Audubon's home, for I knew that long ago it was dismantled and lost. On the way, I passed Trinity Church Cemetery, where a twenty-eight-foot-high Celtic cross marks the burial site of Audubon, a location he picked himself. The bluestone monument was created with funds solicited by the New York Academy of Sciences, in part to successfully forestall the extension of what is now Audubon Avenue. Carvings on the monument include a bust of Audubon and many of the birds and mammals he depicted in his publications. Strangely, the first bird shown is a vulture—on what is essentially a grave site.

The location of the churchyard and the landmark Church of the Intercession was part of Audubon's estate, which extended from 155th Street to 158th Street, and which he purchased in 1841 and named Minniesland. At that time the area was a rock slope containing a wooded clove consisting of great elms and oaks interspersed with dogwoods and tulip trees on the hillside and tall pines nearer the Hudson River. Standing in the west section of the church cemetery at 155th Street, the viewer sees how the elevation and configuration of the land, and the narrowness of Manhattan Island rising northward, deterred the original commissioners from applying the grid farther north. Just to the north of 155th Street, streets curve downward to the Hudson as they do on the east side.

Audubon's home was located on the north side of 155th Street near Riverside Drive, about forty feet below the present surface. After Audubon died, his wife, Lucy, began work as a teacher and sold off parcels of land, which had by that time become known as Audubon Park. In addition to her grandchildren, she taught local children who lived in Audubon Park, one of whom was George Bird Grinnell. Grinnell went on to serve as a naturalist and mineralogist on western expeditions, later writing extensively about western Native Americans and supporting movements to preserve wildlife, as well as generally supporting conservation in the American West. With Theodore Roosevelt, he was a founding member of the Boone and Crocket Club, dedicated to the restoration of America's wildlife (a club that, by the way, met for a time at the American Museum of Natural History). Grinnell created the first Audubon Society, was an organizer of the New York Zoological Society, and was the editor of the influential *Forest and Stream* magazine from 1876 to 1911. So Audubon Park's influence connects the beginning of the conservation and national park movements in the United States and extends far beyond Audubon's folios.   *SH*

Criticism of the plan, as was already the case with earlier attempts, centered on the fact that the commission paid no attention to topography and little to existing roads. Ten open spaces (about 470 acres total) were made part of the plan, including the Grand Parade between Twenty-Third and Thirty-Fourth Streets, and between Third and Seventh Avenues (a huge space!), to be used, as noted by William Bridges, as a place to "assemble, in case of need, the force destined to defend the city," as well as several small parks, which included Manhattan Square—future home, as we have seen, of the American Museum of Natural History. Most of these parks are now gone, and the Grand Parade has been whittled down to Madison Square Park (ever wonder why the succession of basketball/hockey/circus/rock concert arenas there have always been called "the Garden"?). The commissioners wrote, as noted in Bridges's book,

> It may, to many, be a matter of surprise, that so few vacant spaces have been left, and those so small, for the benefit of fresh air, and consequent preservation of health. Certainly, if the City of New York was destined to stand on the side of a small stream, such as the Seine or the Thames, a great number of ample spaces might be needful. But those large arms of the sea which embrace Manhattan Island render its situation, in regard to health and pleasure as well as to convenience of commerce, peculiarly felicitous. When, therefore[,] from the same causes the price of land is so uncommonly great, it seemed proper to admit the principles of oeconomy to greater influence than might, under circumstances of a different kind, have consisted with dictates of prudence and the sense of duty.

Central Park was as yet undreamt of.

The commissioners hired the earlier-mentioned John Randel Jr., who had previously worked for Simeon DeWitt, the surveyor of New York State, to survey the island. Randel and his surveyors eventually measured every inch of Manhattan; in some places the island was so thickly wooded that the going was "impassable without the aid of an axe." Randel created a large topographical map for the commissioners after they had already decided on the gridiron plan, drafting three copies (each measuring 106 inches by 30⁷⁄₁₆ inches), on which he superimposed the "orderly plan of avenues and streets up to 155th." This unimaginative plan, too, failed to take into account the hilly terrain. Robert Augustyn and Paul Cohen note that Clement Clarke Moore (otherwise remembered for *The Night before Christmas*) wrote, "These are men ... who would have cut down the seven hills of Rome." Frederick Law Olmsted, critical of the plan, much later managed to preserve some of the natural topography in his design of Central Park with Calvert Vaux.

Though the plan is sometimes referred to as the "Randel Map," Randel's name is not actually on the published versions. As he began preparing his map for the engravers, he found out that the city "had given the privilege of publishing it to Mr. William Bridges." Bridges was politically connected and was at that time the city surveyor of New York; in 1811 he issued the Commissioners' Plan of 1811 along with a fifty-four-page descriptive pamphlet. Bridges's map contained errors, both because he did not have access to the original papers and because he could not legally render an accurate copy of the Randel Map, which was already on public record. Though Randel drafted another map in 1814, he withdrew it because he didn't want such an accurate map to fall into the hands of the British, with whom the United States was still at war.

Once the city council approved the map and plan, they called for elevations of each street to be drawn, as well as a marker to be placed at each intersection. Between 1811 and 1821, Randel and his workers placed at each intersection a white marble marker measuring three feet nine inches long and engraved with the street's number. Where rocks blocked the way, half-foot-long iron bolts were attached to the rocks. A total of 1,549 marble markers and 98 bolts dotted the landscape of the island. It was now easy to determine how far you were from city hall. For example, Houston (formerly North) Street was about one mile north (twenty blocks equal one mile). At Sixtieth Street you were four miles from city hall. The avenues were numbered and almost equally spaced. Lexington and Madison Avenues were constructed later as a result of the rapid growth on the East Side and the need for more north-south roadways.

But roads weren't enough to complete the concretization of Manhattan Island.

NEW YORK ON RAILS

The Erie Canal opened in 1825, another key signal of New York City's burgeoning status as the commercial hub of the nation. With the canal came the railroads. At first these were feeder lines to the canals, but soon they became interregional modes of transportation of goods and passengers in their own right. As Eric Homberger reports, a horseback journey from New York to Boston took from four to six days in the 1780s, while a stagecoach ride along the Boston Post Road to the same destination fifty years later reduced the duration to a day and a half. But by the 1840s, New York to Boston by rail took about a half a day—not much more than today's journey, despite the arrival of high-speed trains.

Rail lines were also established within the city itself simply to improve public transportation. As commercial interests moved ever northward, and the increasing population in already overcrowded tenements drove northward expansion, better ground transportation became imperative. In just a little over a century, roughly between 1830 and 1940, New York's surface rail, the elevated trains (the "El"), and finally the subway systems were constructed, though the system has always remained a work in progress. Many social reformists (such as the renowned Jacob Riis) saw in such continual expansion of rail lines the solution for New York's perennially overcrowded living conditions for the poor, who could conceivably live in single-family dwellings, ideally privately owned. But in truth, every time a rail line, an El, or finally a subway line was either first built or extended—giving access to farm fields or only lightly urbanized neighborhoods, first on Manhattan and later in the outlying boroughs of the Bronx, Queens, and Brooklyn—usually the opposite happened. Land values jumped with the arrival of railroads in the neighborhood, and valuable land for the most part meant construction of apartment buildings rather than spread-out single-family dwellings. Ironically, though, land values often decreased close to sections of the railroads and Els themselves, where the noisy, smoke-belching locomotives made life much less pleasant.

New York's first railroad was chartered in 1831. The original plan was for the New York and Harlem Railroad to run from Twenty-Third Street up to the Harlem River along a route somewhere between Third and Ninth Avenues; the company chose Fourth Avenue. A trunk line was to be built westward to the Hudson River somewhere between 125th and 129th Streets. There is a convenient geological fault, aptly named the 125th Street Fault, that runs across Manhattan at that point. The ridge of rock forming Morningside Heights and Hamilton Heights is broken there and eroded away along the fault, providing easy, flat access to the Hudson River. That is why the subway emerges from its underground tunnel north of the 116th Street/Columbia University Station and runs outdoors on an arched viaduct. The 125th Street station is a high climb from street level; the next stop north (137th Street/City College of New York) is underground, because the rock formation reemerges to form a high ridge running all the way north to Dyckman Street—where it is again broken by a cross-cutting fault.

A second train extension, south from Twenty-Third Street to Prince Street and also to the Bowery, was granted in 1832. Ground was broken on February 25, 1832, for an extension of the first of New York's railroads, at Thirty-Second Street and Fourth Avenue—which at that time was three miles north of "town." Three major obstructions lay ahead: Murray Hill (between Thirty-Second and Forty-Second Streets), Mount Prospect (between Eighty-Eighth and Ninety-Fifth Streets), and Harlem Creek, farther to the north. The first stage of the project was opened for business on November 11, 1832, running south from Thirty-Second Street. The coaches were horse-powered and remained so at least in central Manhattan over the years that thirteen additional railroad lines were opened. By 1860, these horse railroads were transporting nearly 40 million passengers a year.

Figure 50. Looking north on Amsterdam Avenue to the valley at 125th Street. This valley eroded along New York City's largest fault, which occasionally produces minor seismic activity. Photograph by Sidney Horenstein.

Murray Hill and Mount Prospect were breached by tunnels—still a prodigious engineering feat in the 1830s. Drilling, blasting, and boring their way through, the workmen finished creating an open cut through Murray Hill, between Thirty-Second and Fortieth Streets, in 1837; in 1850 the cut was roofed over. Vestiges of this tunnel remain as a car route along what is now called Park Avenue. The 600-foot-long Yorkville Tunnel through Mount Prospect, also completed in 1837, plus the 658-foot trestle bridge spanning Harlem Creek, completed the railway line. Coal-powered steam locomotives plied the tracks down as far as Thirty-Second Street, where the quieter and far less polluting horses took over. After all, what's a few horse droppings compared to voluminous billows of dark, choking smoke? Locomotives ventured south of Thirty-Second Street along New York's first railroad line only when traffic volume was inordinately high.

The elevated lines were another story. To keep pace with the demands of an ever-expanding population, and after debating the relative virtues of subterranean versus street-level versus high-rise approaches to mass transit (subways were deemed four times as expensive as elevated tracks, and higher-speed train travel at street level was not really an option), the New York Elevated Company began operating the first of the Els, along Ninth Avenue in 1871. Running from the southern tip of Manhattan to Thirtieth Street, the line was extended north to Fifty-Ninth Street by 1876. The clamor to extend it still farther (this included the voice of the newly minted American Museum of Natural History, which was desperate for public transportation to bring visitors to its then-isolated location) soon gained an extension of the Ninth Avenue El uptown, where it reached 180th Street by 1880.

The Second, Third, and Sixth Avenue Els followed hard on the heels of the Ninth Avenue line. According to Homberger, by 1890 New York City "had a mass transit system with greater total mileage than London." But it was already clear by then that the El system was just not enough. The Els hung on for a while and eventually became electrified—and even partly integrated with the newer subway system. The last of them, the Third Avenue El, was torn down piecemeal between 1950 and 1973. But shortly after the turn of the century, the focus of transportation became subways, subways, subways.

The monumental blizzard of 1888 paralyzed New York. Theodore Roosevelt somehow managed to walk to what is now the New York

Historical Society at Central Park West and Seventy-Seventh Street from his home on the East Side. Meanwhile, fifty-mile-an-hour, snow-laden winds sweeping out of New Jersey across the Hudson heaped snow into twenty-five-foot-high drifts, covering street lamps and reaching the second story of some buildings. The blizzard shut down every transit line in the city, preventing people from getting food or heating fuel. Mayor Abram S. Hewitt, a month and half before the blizzard struck, had proposed a subway system that, with its higher speeds and ability to carry more passengers, might well have stayed in service during this wintry emergency. But it would be another decade before his and similar proposals for a subway system were taken seriously.

The first subway opened in 1904, and today's system of 840 miles of track and 469 passenger stations was essentially completed by 1940. The Interborough Rapid Transit (IRT) line was the first to begin operations, with the immediate effect of adding to the population of the northwestern and west Bronx. Brooklyn-Manhattan Transit, a second privately held company, was formed in the 1920s, followed in the 1930s by the Independent line, a system funded solely by the City of New York. The effect was startling: by 1910, New York's population had already reached 4.8 million, and subway connections now extended not only farther uptown but also to the Bronx, Queens, and Brooklyn. The fifth borough, Staten Island, is of course separated by too much water to allow rail connections with Manhattan. To this day, the Staten Island Ferry remains the sole connection between Manhattan and Staten Island. However, the Verrazano-Narrows Bridge, completed in 1964 across the Narrows, connects motor vehicles and New York Marathon runners to Brooklyn, while the Goethals and Outerbridge Crossing Bridges connect Staten Island to New Jersey, with which, at least topographically, every map shows Staten Island is most closely affiliated.

With the expansion of the American population westward, and the consequent rise of the importance of canals and especially railroads carrying freight and passengers from the big cities in the east, New York's prime location as an oceanic natural harbor was beginning to cramp its

Figure 51. The Staten Island Ferry viewed from Battery Park. The ferry, one of the important connections between Manhattan and Staten Island, not only is routinely used by commuters but also is a major tourist attraction. Commuters read newspapers, tourists take pictures. Photograph by Sidney Horenstein.

style. Not only was Manhattan an island, but also Brooklyn and Queens (not to mention the rest of Long Island) were separated by water from the Bronx—and the Hudson River itself was a formidable barrier between New England and eastern New York State and the rest of the country.

Something, of course, had to be done to give New Yorkers greater ease of passage among the boroughs—and to give the city more than purely water connections to the rest of the country. The answer of course was its still-growing system of bridges and tunnels. Easily the most famous of its 2,027 bridges (in itself an amazing statistic!), the Brooklyn Bridge was the longest suspension bridge in the world when completed and the first to use steel cables. It opened in 1883. The Triborough Bridge opened in 1936, connecting Manhattan with the Bronx and Queens—while the famous George Washington Bridge, still the

Figure 52. The walkway of the iconic Brooklyn Bridge, a suspension bridge that opened in 1883. At the time, it was the longest bridge of its type, and it connected what was then the city of Brooklyn with New York. More than 120,000 vehicles, four thousand pedestrians, and thirty-one hundred bicyclists cross the bridge every day. Photograph by Sidney Horenstein.

only bridge connecting Manhattan with New Jersey and points west, was completed in 1931. (The next bridge spanning the Hudson farther north is the Tappan Zee, which takes the New York State Thruway over the Hudson. This bridge was not built until 1952 and is now, as we write, considered due for an expansion or replacement.)

While bridges accommodated mostly vehicular traffic, tunnels carried the day for both railroad and subway connections between the boroughs. By 1906, sixteen tunnels were either completed or in the process of being excavated under the Hudson, Harlem, and East Rivers, which led journalist Arthur B. Reeve to quip that "Manhattan may be described as a body of land surrounded by tunnels."

The engineering art and science of boring through bedrock had by the turn of the century reached a level of sophistication that allowed

QUEENSBORO BRIDGE AND EAST RIVER

I have trekked over all the walkable bridges in New York City, and each possesses its own "feel" and vistas. The Queensboro Bridge remains my favorite, not only because of its amazing steelwork, but also for its spectacular views north and south. Of the major bridges in the city, this one (formerly called the Blackwell's Island Bridge, and now the Ed Koch Queensboro Bridge) is one of three cantilever bridges; the others, the Goethals Bridge and Outerbridge Crossing, are on Staten Island. Unlike many of the other bridges, this one was built as a cantilever type because that made it cheaper and stronger than a suspension bridge and, at the time, allowed for a much longer span. Like all bridges, it possesses a symmetrical outline, but here the details are angular because of the numerous triangles that make up the bridge's structure; it also has an array of eyebolts instead of cables.

Beginning in the mid-1800s, Queens residents and politicians clamored for a bridge connecting midtown Manhattan to Queens. Although there were several false starts, the actual planning process did not get under way until Queens (and Brooklyn, too) agreed to join the City of Greater New York in 1898, in part to gain improvements in public transportation and an adequate water supply. The New York Board of Aldermen (today, the city council, more or less) agreed to build the bridge. Begun in 1902, it was completed in March 1909 and, ever since, has been a subject of movies, songs, art, and books. And for neighborhood kids, like me, it was a compelling playground spanning the East River. This "river" bisects New York City, and the bridge leaps over the river's East and West Channels but has a firm footing on Roosevelt Island in between. The East River—a sixteen-mile tidal strait completely dominated by the tides, and which meets Long Island Sound in the north and Upper Bay in the south—was where the seaport began in Dutch times. The river remained a major waterway for the industrial complex that grew up along its banks. Today, in addition to the bridges, fifteen tunnels (some of them double tunnels) join the two parts of New York City.

For us local kids in Queens, entry to the bridge was at the stubby south tower on Vernon Boulevard. Here a few steps took you through a doorway to a dim space where elevators waited to carry you to the roadway level. During the winter the potbelly stove inside, glowing deep red to keep the elevator operator warm, was a welcoming sight. Once above, you had the choice of taking a trolley car to Manhattan or going up a short stairway to the next level to walk across the bridge—or attempt to climb some of the superstructure (never successfully completed). On the way back, there was no elevator operating on the north side, so you had to descend an iron stair to a catwalk beneath the roadway and walk to the south tower and the elevator. For the nondriving public, this was the link to the hospitals and other city facilities of Welfare Island (originally Blackwell's Island, now Roosevelt Island), other than the ferry at Eighty-Sixth Street in Manhattan.

Today, the trolley tracks are gone, replaced by vehicular lanes, and all the "welfare" institutions are gone except for a hospital, which will eventually become a major campus for applied science. Access to the hospitals was through what was then called the "upside-down building," because the entrance to the building was at the top. This structure on the north side of the bridge (demolished now and replaced by the present-day tram station) also served as a storehouse for the hospitals. Trolleys, too, would stop there and then continue to Second Avenue between Fifty-Ninth Street and Sixtieth Street, to the underground station. Access from the station to the surface was through art nouveau–style kiosks, one of which is now an information booth on Roosevelt Island and the other the entrance to the Brooklyn Children's Museum. It all changed when the Roosevelt Island Bridge from Queens opened in 1955. The ferry ceased operation, but the trolley persisted for a while; its last run was on April 7, 1957.   *SH* this rampant tunnel construction. Tunnels were dug in order to build subway lines under the streets of New York; others were designated as water and sewage tunnels (see the next chapter). But they were also dug through the variable matrices of solid bedrock or looser unconsolidated sediments under the rivers.

The history of railroading in and around New York City is in itself labyrinthine, if only because all major railroads in the United States wanted to have a terminus in, or at least near, the city. In 1825, John Stevens was the first person to build a successful steam railroad engine in the United States. One of the first major railroads to serve New York City from outside the city was the New York and Erie Railroad, which ran from Dunkirk at Lake Erie, across central New York, and then down the western side of the Hudson to Piermont, New York. The famed milelong pier at Piermont still stands (and is visible to the south of the Tappan Zee Bridge), which was built to accommodate ferries that would tie up, take on a train's cargo, and transfer it the rest of the way down to New York—or, of course, vice versa. Built in 1841, the pier is where it is (as so many things are) because of a unique combination of local geology and hydrology: there is a fault-produced gap in the forbidding diabase cliffs of the Palisades at Piermont, allowing trains to pass through from the west; and the Hudson is still fully tidal there, affording full access to the sea and, especially, to New York City.

By 1887 only the New York Central Railroad entered Manhattan—over the recently reopened bridge at Spuyten Duyvil along the Hudson, and also farther east over a bridge that led directly to tracks down today's Park Avenue, terminating at Grand Central Station. All the other railroads still terminated on the New Jersey side of the Hudson or on the Long Island side of the East River. The only bridge (there were no tunnels yet) crossing the Hudson was at Poughkeepsie, seventy-five miles to the north. The absence of bridges and tunnels lengthened passenger journeys and raised transportation costs.

On July 30, 1887, the New York and Long Island Railroad Company was incorporated. William Steinway, the famous piano manufacturer, had a great interest in the affairs of New York City, not least because he had established a factory in Queens (moving from New York in part to avoid ongoing labor problems) in 1873, where he planned to produce twenty-five hundred instruments a year. He was also into real estate. A promoter of Queens, which was not yet part of New York City, and its potential, Steinway became the railroad's largest stockholder in July 1891, a role that fit with his other commercial efforts. He established a village around his Queens factory to house his workers and keep outside labor agitators at bay. He built houses, financed a church, maintained the streets, and constructed a firehouse, a library, and schools. He encouraged the Long Island Railway to extend its lines to his factory doors, promoted the plan to build the Queensboro Bridge (completed in 1909), and owned most of the trolley lines in western Queens by 1893. He was quite an entrepreneur, opening New York's first bathing beach and amusement park, called the North Beach Amusement Park (which is now partly covered by LaGuardia Airport) and attracting some fifty thousand New Yorkers each summer weekend.

The railroad's trolley tunnel slated to be drilled under the East River was to bear Steinway's name. Steinway's ambitions were stymied for a while when, on December 28, 1892, an accidental dynamite explosion killed five workers, causing considerable property damage and resulting in legal claims that drained the finances of the company and halted construction of the project for ten years. In 1902 August Belmont, whose name is memorialized at the Belmont Park Racetrack in Elmont on Long Island, bought the New York and Long Island Railroad as part of his plan to dominate mass transit in Queens. Engineering techniques had improved dramatically since 1892, and the Steinway Tunnel opened for business on September 24, 1907.

The Steinway Tunnel consisted of two tubes constructed from four shafts, one of which was located on Man-o'-War Reef just south of Blackwell's Island (now Roosevelt Island) in the middle of the East River. The first trolley made it through the tunnel in four minutes, eliminating the half-hour ferry ride; the line connected to the New

Figure 53. The churchyard of Trinity Church at the western foot of Wall Street. This yard contains the graves and headstones of well-known colonial and Revolutionary War individuals such as Alexander Hamilton, Robert Fulton, William Bradford, and Francis Lewis. Burials no longer take place here, since the city banned them below Canal Street in 1822. However, the churchyard is an educational resource for people interested in early headstone carvings and the study of stone decay. Photograph by Sidney Horenstein.

York and Queens County Railway, a forty-one-mile network of trolleys that essentially started at Grand Central Station and the Forty-Second Street IRT. This was the beginning of extraordinary changes in Queens, a rural area that housed just 4 percent (152,999 people) of the city's population in 1900. Soon after opening, the tunnel became part of the IRT Flushing Line (number 7).

The age of tunneling under the rivers for railroads was born. The Hudson tubes, opened in 1908 and 1909, finally cracked the problem of bringing trains directly from New Jersey to the city, where they now stopped at the grand marble homage to Rome's Caracalla Baths, Pennsylvania Station. (The station was completed in 1910 and wantonly torn down in 1963.) Trains would then traverse Manhattan via tunnel, heading for Long Island or, instead, turning north and coursing along a prodigious railroad bridge paralleling the Triborough Bridge and heading for New Haven and Boston. New York was finally connected.

By 1910 many New York City neighborhoods already showed their essential character: Harlem's great renaissance began then, groceries in Little Italy sprang up (some of which remain open today), and long lists of desirable Upper West Side apartments for rent today were renting then. Downtown today, the Tenement Museum on the Lower East Side draws on oral histories told by an elderly generation who once lived there. Greenwich Village was already attracting its bohemian crowd, and Wall Street looked much as it does today. Uptown, the palaces of patriarchs dating back to the twilight of the robber baron era live on as public galleries, such as the Cooper-Hewitt, National Design Museum, at 2 East Ninety-First Street, and the former residence of steel magnate Henry Clay Frick. In Queens, Ozone Park saw its first planned housing development, and in the Bronx the Zoological Gardens and Botanical Garden were popular destinations for city residents.

CHAPTER FIVE

Fouling, and Cleaning, the Nest

Like most cities, New York has a history of pollution that is long and dirty. We are standing on the Hudson River docks in Chelsea trying to get a cheap lunch of oysters freshly dredged from the Hudson. Today, a probing search might turn up piles of empty shells; but in the 1880s, oyster barges lined these shores and oyster shacks stood at street corners. We have already encountered Peter Kalm, the Swedish botanist and early student of Linnaeus, who came in 1748 to check up on immigrant settlers in Glassboro and Swedesboro, New Jersey, and to look into exploiting the botanical resources of New York, New Jersey, Pennsylvania, and Canada for cultivation back home. In his account *Travels in North America,* Kalm wrote about seeing oysters gathered from near New York City in great number—some of them over a foot long—to roast on the half-shell over a burning log. At one time, the waterways around New York contained about a quarter million acres of oysters. Not today.

Now, even if you could find an oyster here, you could be taking your life into your hands if you ate it, although the Hudson is actually cleaner today than it has been anytime in the last century. (Ironically, cleaning New York's harbor waterways has allowed the return of destructive, boring marine invertebrates, now busily engaged in destroying wharf

Figure 54. The Chelsea docks. Located along the Hudson River south of Twenty-Third Street, and once the main port for luxury vessels, the docks have been converted to a major sports venue and marina. It is here that the *Titanic* was to dock, and where the *Carpathia* brought some of the survivors. Photograph by Sidney Horenstein.

pilings that had, in the polluted past, no need for protective creosote!) At times, the city has paid next to no heed to sanitation concerns and the concomitant spread of disease, but at others it has enforced cleanup laws and taken specific steps such as the 1822 ban on human burial south of Canal Street. At that time, Trinity Church, which was located at the foot of Wall Street, bought land from James Audubon for a cemetery at 155th Street (where Audubon himself is now buried).

DON'T DRINK THE WATER

Manhattan's ideal location in a safe and accessible marine setting brought the blessings of easy transport and trade; but as we have seen, the lack of freshwater was always a nightmare of sanitation and supply

Figure 55. Decayed pier pilings along the Hudson River. This pier was once part of a shipping industry that has since moved to other parts of the harbor. The decay is enhanced by the cleaner condition of the Hudson River, which has allowed marine organisms that feed off the pilings to repopulate them. Photograph by Sidney Horenstein.

challenges. After Henry Hudson continued up his namesake river while searching for the elusive northwest passage to Asia, the first settlers knew that, because it had proven salty, the Hudson would not be a source of potable water. Yet they started their community at the southern tip of Manhattan Island with evident confidence that there would be enough water derived from land sources. Using techniques learned in Europe, they tapped small springs and brooks that ran from the base of the hills which then stood on the island, supplemented by rain runoff collected in roof cisterns and probably by shallow wells. This seemed adequate for several decades, until the need grew to consider a formal plan to dig deeper wells, one that was drawn up in 1658. The bottles of "designer water" we see everywhere today in the hands of runners and strollers echo the scorn consumers had for the waters of old Manhattan.

As the population and city grew, additional wells were provided, but complaints about the quality of the water increased as poor sanitation—partly the result of the large number of pigs roaming the streets—led to an ample amount of pollutants. In addition, overdrawing the well system contaminated the aquifer with salt water, causing residents to seek freshwater elsewhere. This marks the beginning of the city's quest to support itself outside its boundaries.

The Dutch surrendered New Amsterdam at least in part because they ran out of water inside their fort, where they had no well and only a limited supply of water stored in kegs. New York City was never destined to support itself with local water—even if it had kept the water clean. Drainage and filling in of ponds, diversion of streams, and pollution of groundwater—plus population growth—all exacerbated the water shortage and led to the development of the elaborate system of dams, tunnels, and reservoirs that are the hallmarks of the present water supply system. Water supply—so essential to human life, yet requiring so many demands on and changes to the natural environment—is an especially graphic example of the impact the concrete jungle has on environments as far removed as 120 miles from the city itself. And there is truly nothing new here, as the Roman aqueducts were merely early versions of New York's extensive water supply system.

As in ancient Rome, nothing was ever accomplished in New York without a political angle. After the city's start on a municipal water system, developed in 1774 by Christopher Colles, was destroyed during the Revolutionary War, the first major waterworks plan came from Aaron Burr, whose duplicity rivaled that of any modern lobbyist. In his quest to create a private bank to compete with that of his political enemy Alexander Hamilton, Burr managed to divert public money for a municipal water system into his Manhattan Company, whose investors purchased shares of stock. The company controlled large sums of money—along with water delivery. In the twentieth century, a remnant and memory of the bank survived in the Chase Manhattan name, now truncated to Chase. The Manhattan Company built a company-owned reservoir in

1799 on Chambers Street, where the Hall of Records stands today, and it put up an iron holding tank that remained on Centre Street until the early 1900s. The actual well supplying water to the reservoir was discovered during work on a city office building at Duane and Reade Streets in 1926, and all the wooden pipes occasionally dug up during construction projects come from this era. Pine logs harvested in the Adirondacks and measuring up to fifteen feet long (most were shorter) were hollowed out (six- to nine-inch holes were bored in logs with a one-foot diameter) and linked to distribute water through about twenty-four miles of wooden mains. Inevitably, the wood rotted and complaints about the foul-tasting water became common, so by 1825 the company began slowly replacing the wooden pipes with metal. Even so, only wealthy people could afford the water, although many poachers also tapped the supply. And, as we have seen, for a time carters drew barrels of water from Collect Pond and the nearby Tea Water Gardens to deliver around the city for drinking, but that trade was killed by pollution. Collect Pond's decline from its status as Fresh Water Pond to a sewer of preindustrial tannery waste and butcher's offal ultimately produced an unbearable stench that had to be quenched. Collect Pond, along with every other watercourse downtown, was filled in by 1811 (filling began in 1803), a process known to engineers as "rectifying the surface," and today no open water remains south of Central Park. In accordance with the original Manhattan Company charter, the water supply was also used to fight fires. A fireman would remove the soil cover to expose a wooden pipe and, with an auger, bore a hole in it, allowing access to the water by hose. After the fire, a mallet and a wood plug restored the integrity of the pipe—hence the term *fireplug*, an old term for "fire hydrant."

REGIONAL EXPLOITATION: THE CROTON WATER SYSTEM AND BEYOND

In spite of the Manhattan Company's charter allowing fire brigades to draw water from their pipes, the Great Fire of 1835 devastated the city.

Before that, pestilential, communicable, and waterborne diseases, including smallpox, cholera, yellow fever, and typhoid, were rampant. Devastating yellow fever epidemics hit the city in 1819 and again in 1833, while a cholera epidemic in 1832 killed 3,513 of the city's 250,000 residents and 100,000 more fled the city. Further troubles came from widespread economic depression starting in 1837. With the electorate's approval in 1835 and a loan of $2.5 million, the city fathers looked to their future and launched plans to tap the Croton River despite the hardships of the depression.

After a few false starts, the job fell to John Jervis, who was self-taught like all engineers except for the military men trained at West Point and specialists from Europe. Jervis had learned his trade by working on the Erie Canal. After voters approved a new water system in 1835, the city hired Jervis in late 1836 to create a new water supply system that would tap the Croton River, implementing and improving the earlier conceived plan. The following year, Jervis sent a crew of nearly four thousand immigrant laborers to begin work on the Croton Dam in what is now Westchester County. It took five years to finish, cost the colossal sum of $12 million, and laid the foundations of an abiding tension between the city and the people living near its rural sources of water. The local landowners and farmers whose four hundred acres were condemned to create the five-mile lake, the dam, and the aqueduct naturally protested. But New York State law set a still-honored precedent and found in favor of the right of New York City to condemn property in the just cause of an adequate water supply, so long as landholders received fair compensation. Those rights are variously protested, contested, and detested to this day. Perhaps the nineteenth-century nickname for the familiar city-dwelling cockroach—Croton bugs—reflects that hostility, since legend has it that the European immigrants who supplied the majority of labor on the dam unwittingly imported the roaches on their journey to the New World.

On June 22, 1842, the first Croton water flowed to the end of its forty-one-mile journey from upstate, across the Harlem River, into the

Figure 56. The original Croton Dam. Completed in 1842, the dam was engineered by John Jervis, one of the great early American self-taught engineers. The ogive curve of the dam face and the stilling basin in front of it reduced the force of the falling water and prevented erosion along the toe of the dam. Jervis was first to use this design in the United States. The original Croton Dam reservoir delivered wholesome water for the first time to the distribution reservoir at Forty-Second Street, now the site of the New York Public Library, a distance of forty-one miles from the dam. From *Illustrations of the Croton Aqueduct,* by F. B. Tower (Putnam: 1843).

reservoir site (now occupied by the Great Lawn of Central Park), and thence to a grand distribution center (built on the model of an Egyptian temple), where the Forty-Second Street Public Library now stands. The reservoir was filled by July 4, 1842, and its waters became available for public use on October 14, 1842. (The fact that the adjoining grounds of today's Bryant Park were occupied by the Great Exhibition Hall—New York's version of London's Crystal Palace—which was built in 1853 and destroyed by fire in 1858, explains why two prime pieces of adjoining real estate were available for the library and its grounds when work started in 1897.) Where the Croton pipeline descended to cross the Harlem River, the gravity-feed system laid over pontoons had to be vented, and

Figure 57. Durable Devonian-age bluestone from the Catskill Mountains, where at least a hundred quarries operated during the late nineteenth and early twentieth centuries. This resistant sandstone was used not only for sidewalks, such as the one around the New York Public Library at Forty-Second Street, but also for a variety of building projects. Photograph by Sidney Horenstein.

passersby could see a great jet of water spouting to relieve pressure, until the High Bridge was completed in 1848. Although this landmark icon, the oldest remaining bridge built in New York City, looks like a faithful copy of a Roman aqueduct, its construction represents the most exacting modern engineering standards of the day. The new bridge gave its name to Highbridge Park in Manhattan, as well as to the facing neighborhood in the Bronx, and was an attraction as a tourist destination—leading to the creation of two amusement parks north of the bridge. In 1928 five of the original fifteen arches were replaced with a central steel span to allow larger vessels to pass under it. The rest of the stone arches remain, mainly on the Bronx side of the river. The bridge was never built for traffic, but the top of the bridge was used by pedestrians, beginning shortly after it

Figure 58. High Bridge, which crosses the Harlem River, as conceived in 1842 and completed in 1848. The bridge is part of the aqueduct system that brought water from the Croton Dam, across the Harlem River, to the distribution reservoir at Forty-Second Street. Today, five of the arches have been replaced with a steel arch to improve navigation in the river. It is the oldest extant bridge in New York City, and its restoration is nearly complete. From *Illustrations of the Croton Aqueduct,* by F.B. Tower (Putnam: 1843).

opened until 1960, when it was closed after delinquents dropped rocks from there onto Circle Line boats passing below.

The completed Croton system could deliver 90 million gallons a day, which should have provided plentiful water for decades into the future, but demand outpaced supply almost immediately. As population and industry increased, and as water closets and new plumbing devices such as indoor toilets and sinks with faucets were introduced, the city saw an exponential increase in water usage. With nothing to carry off locally disposed wastewater, the groundwater level rose many feet by the 1850s, and the Croton Aqueduct Department was obliged to remedy matters with seventy miles of sewers laid between 1850 and 1855. The search was already on for new water sources when a series of droughts intensified the problem.

The interim solution was to expand the Croton system by adding new reservoirs by tapping the tributaries of the Croton drainage basin,

THE HIGH BRIDGE

In the 1950s, while attending school in the Bronx, I walked across the High Bridge several times, enjoying the views, north and south, of the Harlem River Valley, the Harlem Drive in Manhattan, and the railroads near the Bronx shore. The Major Deegan Expressway was nearing completion. I also imagined what the views were like when, as it is said, Edgar Allen Poe walked the bridge for consolation while living nearby in the Fordham section. High Bridge was closed to pedestrians in the 1960s as a result of vandalism and its use as a hangout for delinquent youths, who sometimes threw objects off the bridge onto passing Circle Line vessels and injured passengers.

The last time I walked the bridge was in 2010, during the celebration marking a start to restoring the bridge. After its completion, the public and I will have access to the bridge once again. And they will also be able to visit the now beautifully restored cottage where Poe lived when he wrote "The Cask of Amontillado," "The Bells," and "Annabel Lee." The view from the bridge also reveals one important aspect of New York's geology: the composition of its bedrock is unequal in susceptibility to weathering and erosion. The Harlem River, actually a tidal strait, is sandwiched between two resistant rock formations, the 565-million-year-old Manhattan schist on the Manhattan side and the 1.1-billion-year-old Fordham gneiss in the Bronx, the oldest rock in our region. Both of these formations are more resistant than the 500-million-year-old Inwood marble at the foundation of the valley itself. So, through the millennia, erosion has picked away at the marble more than at the schist and gneiss, creating the valley. And looking far north from the bridge, you'll see the wooded Van Cortlandt Park ridge in the distance, underlain by the Yonkers gneiss, another resistant rock formation, one that is found only in the Bronx and farther north in Westchester County.

The High Bridge is the oldest extant bridge in New York City; it's listed on the National Register of Historic Places and is a New York

City Landmark. It carried New York's first dependable water supply system, from the Croton River across what until then had been a barrier presented by the Harlem River. In 1835, as a result of disease and earlier disastrous fires, the voting citizens approved funds for construction of the system (the Croton Aqueduct or Old Croton Aqueduct) to bring pure and abundant water to the city. The project's design was awarded to a West Point–trained engineer, Major David Bates Douglass, who was fired and replaced by John Bloomfield Jervis after eighteen months. Jervis gained his skills by working on the construction of the Erie Canal; the High Bridge and the Old Croton Aqueduct water supply system stand as testimony to his considerable abilities.

Douglass was involved in many projects that improved New York's position economically and socially. In addition to engineering the original Croton Dam and water supply system, he was the chief engineer for the Morris Canal (1829–1832), which brought coal and other goods to New York from Pennsylvania; he surveyed the route of the Brooklyn and Jamaica Railroad on Long Island (1832–1834); and he was designer and engineer of Greenwood Cemetery in Brooklyn (1839–1840), now designated a National Historic Landmark. *SH*

but by 1880 plans were afoot for the New Croton Dam and Aqueduct System. The Aqueduct Commission began acquiring the twenty-square-mile area (in Westchester County, north of New York City) needed to build the new dam and reservoir. The condemnation process literally uprooted the dead, as well as infuriated the living, as fifteen hundred bodies were moved from six cemeteries before four hundred farms and four towns, with all their churches, schools, shops, and mills, were drowned. The excavation began in 1892 and was completed in 1906, creating a dam heralded as the second-largest hand-hewn masonry structure in the land and the tallest dam in the world.

Meanwhile, in the years after 1842, not all New Yorkers enjoyed access to the new supply of potable water. On the Lower East Side,

Figure 59. New Croton Dam during a drought. Located in Westchester County about twenty-two miles north of the New York City limit, it was the world's tallest dam when it was completed in 1906. The main reservoir supplies 10 percent of New York City's needs from numerous interconnected reservoirs forming the new Croton system. That system, together with systems that bring water from the Catskills and the Delaware River tributaries, supplies 1.5 billion gallons daily. Like the other systems, New Croton Dam is subject to occasional drought, but because the others are in a different climatic regime a drought rarely affects all of them at once. Photograph by Sidney Horenstein.

tenement owners avoided connecting to the Croton water system with the argument that the cost would drive them out of business. As a result, New York had the highest waterborne-disease mortality rates in the United States and western Europe. Records from 1900 show there were dwellers who relied on well water, including those living in shanties around Columbus Circle, and there were still outhouses in the backyards of tenements into the 1930s. At last, in 1904, the law required all buildings to connect to Croton water, but it was the introduction of chlorination following Jersey City's initial example in 1908 that caused waterborne diseases to fall rapidly. This happened only after a court battle pitting concerned citizens against a city's right to chlorinate its water supply. That right was upheld as a safeguard to public health, an action that paved the way for the chlorination of drinking water supplies throughout America.

Even as work progressed on the New Croton Dam, it was clear the city would need yet more water, especially after 1898, when Brooklyn, Queens, and Staten Island joined Manhattan and the Bronx as the City of Greater New York. In 1905 the New York State legislature created the New York City Board of Water Supply and granted it the right to purchase watershed land in the Catskill Mountains, condemn villages, and dam streams and rivers for new reservoirs. By 1915 the Ashokan Reservoir and Catskill Aqueduct were completed, and full development of the Catskill system, composed of the Schoharie Reservoir and Shandaken Tunnel, was finished by 1928.

But as early as the 1920s, J. Waldo Smith, chief engineer of the board of water supply, warned that the burgeoning city's water demand would surely outstrip supply by 1932 unless new sources were added. Once again the city mounted an even more far-flung search and another bitterly fought series of encroachments and condemnations as it zeroed in on the Delaware River basin. The river's east and west branches wound through deep valleys more than one hundred miles to the northwest of the city, in Delaware County, New York, to converge in Hancock. From there, the river defines the boundaries of New York, Pennsylvania, New Jersey, and Delaware as it flows 330 miles south to debouch at Delaware Bay.

In 1923, the New York State legislature created the Delaware River Treaty Commission, never ratified by Pennsylvania and New Jersey, with the federal government as advisor, in search of a plan that might serve as a treaty, a conservation manifesto, and a development strategy: the Tri-States Delaware River Commission. A treaty was approved but immediately ran into politically motivated obstacles. It took three U.S. Supreme Court decisions (in 1931, 1954, and 1961) to assert the claim that since most of the water originated in New York State, New York's claim took primacy over the objections of its rival claimants.

The titanic work of digging more than 150 miles of tunnels began. The tunnel lengths ranged from 83.8 miles for the Delaware Aqueduct connecting the Rondout and Hillview Reservoirs, to the 44-mile West Delaware Tunnel, to the 25-mile East Delaware Tunnel, and to the 6-mile Neversink Tunnel. On March 24, 1937, Mayor Fiorello La Guardia officiated at what turned into an explosive kickoff of construction in Ulster County when the opening blast he ceremoniously detonated to start the tunnel near Lake Minnewaska belched a cascade of rock shards upon the spectators. Although the pace of progress was unparalleled, and the aqueduct was mostly done by 1941, World War II siphoned off the manpower and materials needed to finish the job until 1945. Even so, the impounded waters of the Neversink River and Rondout Creek did not flow through the aqueduct until 1954, when the Neversink and Pepacton Reservoirs were completed.

Further development and expansion continue unabated into the present, but through foresight and princely public investment, New York City had by the early twentieth century secured a wide network of resources with the strategic vision of a modern Roman state. The mountainous location of many reservoirs permits a 95-percent-gravity-feed system, and the low energy costs for pumping are the envy of all.

Part of the Croton system, the Jerome Park Reservoir was named after the Jerome family whose daughter Jenny married a Briton and became the mother of Winston Churchill. The original plan called for a gigantic reservoir in the Bronx stretching from Kingsbridge Road up to the Mosholu Parkway, but the pressing demand for even more water prompted plans to exploit the Catskill region even before the Jerome Park project was completed. So, while the western half of the dig indeed became a reservoir, the eastern half wasn't finished, because the Catskill Aqueduct made it unnecessary. The latter remained a huge depression, which now accommodates Lehman College, the Bronx High

Figure 60. Rooftop water tanks, a vital part of the water supply system of New York. Although water flows to the city by gravity, numerous pressure-regulating valves in the water mains allow water to rise to the tops of six-story buildings. Taller buildings require pumping. Water in the tanks, which are minireservoirs, flows down to residences by gravity. Most of the tanks are constructed of long-lasting redwood, but steel tanks are also in use. Photograph by Sidney Horenstein.

School of Science, Kingsbridge Armory, and the rail yards of the Independent subway line.

The latest installment of the Jerome Park story takes us to another excavation, this one in Van Cortlandt Park. The golf course in the latter park—which opened in 1895, the oldest public golf course in the country—was temporarily torn up to accommodate a water filtration plant that had originally been destined to be built over the Jerome Park Reservoir. In the face of strong community opposition, the city agreed to install a subterranean system and restored the golf course, open again for its second century as a popular and often crowded public recreation site.

THE SEWAGE AND GARBAGE OF GOTHAM

When the population of Manhattan was fifteen hundred in 1664, sewage and other waste was disposed of in pits or open dumps as well as along the shoreline, where new land was created.

The population had mushroomed to three hundred thousand by 1830, and the primary nineteenth-century garbage challenge was well entrenched: how to get rid of all the dead animals. As the century wore on, the indigenous garbage-collectors—the roaming flocks of goats and pigs—made room for ever-greater numbers of horses, each of whom mired the roads with quarts of urine and some ten to twenty-four pounds of manure a day. Until the motorcar's ascendancy, between one hundred thousand and two hundred thousand horses at a time served and lived in the city. It is sometimes said that the elevated stoops of New York City brownstones were designed to give their owners access to higher floors above the streets' stench and filthy runoff, which tended to percolate into the ground floor of residences. If so, they afforded little escape from the pathogen-laden dust of dry, pulverized horse droppings that the wind lifted from deep between the joints of cobbled paving stones and drove indoors. When the horses died, usually of overwork after a lifespan of only about two and a half years, their carcasses, weighing a half ton or more, rotted in place until decomposition left the bones exposed for removal. Further muck came from the citizens' night soil and garbage, routinely tossed out in the streets. Between the debris and snow, a winter's accumulation could raise street level from two to six feet. At last the police department was assigned responsibility for street cleaning in 1872, and a street-cleaning department was formed in 1881.

Even garbage has its brighter side, though: the profit from a variety of animal components. Bone could be worked into buttons and fancy imitation ivory goods, charred and used to filter raw cane juice for sugar, burnt to make a useful dye called bone black, or ground into fertilizer. Hides went to the tannery, while flesh and marrow rendered up their fats to the soap and candle industries; what remained, the pigs would eat. All this messy industry could be found huddled at the north-

Figure 61. A New York City Department of Parks garbage pickup in Isham Park, northern Manhattan. While the Parks Department services the city parks, the Department of Sanitation serves residential structures, and businesses utilize private carting companies. Photograph by Sidney Horenstein.

ern edge of the city, and reports document a large manure field close to the Forty-Second Street reservoir. Before the cholera outbreak of 1849 raised an outcry that led to a ban on soapmaking, open manure pits, and the boiling of livestock blood, entrails, and bones, many such enterprises stood cheek by jowl with the slaughterhouses, livery stables, and dairy barns that the city couldn't run without. But a Boss Tweed–like precursor named Alfred W. White, who was in charge of public health at the time, shut down the worst abuses. He applied a canny mix of legal, political, and business manipulation to clean up the city, initiated a massive pig-clubbing campaign to dispose of feral and some privately owned pigs (some of which were shipped off to the less populated area of the city)—and enriched his cronies, along with himself as a silent partner, in the offshore carting and disposal business he devised to serve the city. The initial destination for the city's garbage, manure, offal, and carcasses—South Brother Island in the East River—was only a half mile from the Bronx shoreline, noxiously close to the sensitive noses of wealthy country estate owners, including the tobacco magnates William H. Leggett and Jacob Lorillard. The adjacent Queens County villages quickly joined forces with their powerful Bronx neighbors and prevailed upon the Queens County Supreme Court to close down the operation.

BARREN ISLAND

Next came the saga of a much-abused portion of Jamaica Bay, itself now part of the extensive federal Gateway National Recreation Area. Today Barren Island is neither, given its greening as a result of a twenty-first-century restoration and a hundred-year accumulation of landfill (consolidating what began as thirty-three islands) spread over thirty acres of upland and seventy acres of salt marsh. In tribute to its wilderness beginnings, the Dutch had named it Bear Island and built tidally driven grain mills along its shore; today walkers along the National Park Service's Millstone Trail can find a last millstone from the era still there. The old harbor named Dead Horse Bay is the more accurate reminder of the island's fate, beginning in the late 1850s, as the destination for all animal refuse and, beginning in the 1890s, for most of the other garbage in New York City. Soon the island became a garbage goldmine. Factories were built to extract fish oils from the rich menhaden catch. By 1860, Barren Island had become one of the world's largest offal-processing sites. The pressings from menhaden, called barrow, were mixed with other animal flesh, ground bone, the waste from sugar manufacturing, and imported Peruvian guano to produce fertilizer that was shipped off to destinations as far afield as vineyards in the Rhine Valley. One factory built in 1878 took in thirty dead horses a day to produce two thousand tons of fertilizer a year. Between 1859 and 1934, there were as many as eight such factories in operation at once, amounting to a total of more than twenty-five all told.

A newspaper article noted in 1912 that the three refuse disposal plants on Barren Island were among the largest of their kind in the world. By then, the menhaden trade was long over, owing to economic depression and the depletion of the menhaden population in the early 1890s. Later, glycerin from the rendered trimmings from butchered animals went into the nitroglycerin used by the Allied forces during World War I. After the war, Floyd Bennett Field, New York City's first airport, derived an advantage from the work done by long years of garbage disposal. In 1926 the city finished the job by filling the remaining salt marshes to connect Dead Horse Bay and the rest of Barren Island to Brooklyn, creating runways that would be used by famed aviators. These included Amelia Earhart, Howard Hughes, and Douglas "Wrong Way" Corrigan, who flummoxed air officials by purposely making the nonstop flight to Ireland he had repeatedly been denied, rather than the trip to California he was registered for.

More modern facilities supplanted the Barren Island garbage dumps, and the dumps were closed in 1918, although one last glue factory there continued to process horses until the early 1930s. Then, in the mid-1930s, with the horse-drawn era coming to a close, the days of rendering plants gave way to the early days of aeronautics. With a stroke of his pen, political giant and New York City parks commissioner Robert Moses condemned Barren Island, granting the remaining handful of residents just a month to clear out before bulldozers razed their cottages and began construction of the Marine Park Bridge that would give access to the new airfield. Today, a stroll along the western shore of the grassy, overgrown tract that was for a time an airport yields glimpses of broken bottles and crockery that may date back to the eighteenth century, along with seamed nylon stockings, cold cream jars, and leather shoes of more recent vintage. But the surrounding waters are so fully recovered from the deadly pollution of the last century that it's safe enough for scuba divers leaving Brooklyn's Barren Island Marina in sportfishing boats to dive for lobster in Rockaway Inlet among the wrecked ships peppering the seafloor.

A CASE STUDY IN POLLUTION: NEWTOWN CREEK

The story of Newtown Creek is a concrete-jungle case study in the rich profits and enduring environmental losses that define America's industrial revolution. The creek, a 3.5-mile-long tributary of the East River, forms the boundary between Brooklyn and Queens. Today much of the creek's marshland is buried, as is part of the creek itself, and it no longer receives natural freshwater surface flows. The Dutch who first surveyed it in 1613–1614 acquired it from the Maspet tribe and cultivated the surrounding land. The local aristocrats built summer homes alongside its marshy banks, where they enjoyed the fishing and hunting. During the American Revolution, the British general Warren was headquartered here, and later DeWitt Clinton's wife inherited the Sackett house, built in 1725 on Newtown Creek, where Clinton planned much of the ambitious Erie Canal project.

The little village of Newtown began in the Queens neighborhood now known as Elmhurst and soon became famous for the apple that originated on the estate of Gershom Moore early in the eighteenth century: the Newtown pippin. At once crisp and tender, it's a green-skinned, yellow-fleshed, sweet-and-tart apple with a citruslike aroma, whose fame spread far enough that it appeared on the 1807 Horticultural Society of London's "Select List" of apples. Thomas Jefferson praised Newtown pippins, and Benjamin Franklin imported them by the barrel in 1759 while living in London. When young Queen Victoria first tasted them in 1838, she responded by lifting the import tax on apples. The mother tree is thought to have succumbed to decades of excessive cuttings in 1807, but on Sunday, April 7, 2002, the First Presbyterian Church of Newtown at Queens Boulevard and Fifty-Fourth Street celebrated its 350th anniversary by planting two young Newtown pippin trees on its front lawn.

Figure 62. Newtown Creek. This 3.5-mile waterway forms the border between Brooklyn and Queens. A major industrial center was built on the banks of this once pristine waterway before it became part of New York City. Oil, copper, paint, and rendering plants all contributed to the creek's extreme pollution, which still dominates its sediments. Below the surface lies a fifty-acre plume of leaked oil containing 17–30 million gallons of petroleum. The creek and part of the adjacent area have been designated as a Superfund site. Photograph by Sidney Horenstein.

Early in the nineteenth century, farmers barged their cabbages and other Long Island produce down the creek to market, little dreaming that a century later the modest waterway would rival the entire Mississippi River in shipping tonnage. The petroleum industry was mostly to blame. New York City offered an abundance of manpower, a superb port, and a prime location for foreign sales. When the first commercially successful oil well, drilled by Edwin Drake in Pennsylvania, kicked off the American petroleum industry in 1859, New York City

Figure 63. Newtown pippins. This light-green apple variety originated as a chance seedling on an estate in the village of Newtown (now called Elmhurst) in the late seventeenth or early eighteenth century. When Thomas Jefferson was in Paris, he wrote, "They have no apple here to compare with our Newtown Pippin." Photograph by Sidney Horenstein.

already had the energy edge with its several plants that distilled illuminating oil from coal, a process similar to petroleum refining. During the 1860s, petroleum refiners arrived in force, since New York was already the leading manufacturing city, with some one hundred thousand workers in five thousand factories. By 1900 its industrial establishments had increased sevenfold.

At first, New York's refineries made just one or two products at random locations along the waterfront. Robert Chesebrough, for example, hit upon petroleum jelly, a by-product of his lubricating-oil business, which he marketed as Vaseline. In 1860, there was only one refinery in Queens, one in Manhattan, and six in Brooklyn. But demand for kerosene both in the United States and abroad expanded production to a full range of refined petroleum products that required large tracts of land for storage, pipelines for transmitting petroleum to refineries, and processing. John D. Rockefeller decided to concentrate the operations of the Standard Oil Company along Newtown Creek mainly because there were refineries there already. The vicinity of Newtown Creek became the undisputed center of petroleum refining in New York.

Beginning in 1872, Rockefeller absorbed or bought a controlling interest in distillers one by one. By the 1880s he controlled over one hundred stills, employing two thousand people and processing 3 million gallons of crude oil each week. Pastoral land adjacent to Newtown Creek was transformed into a vast interconnected complex of wharves, stills, tanks, and pipelines. The refineries anchored a growing number of related industries: companies that constructed stills and tanks located nearby; varnish and paint manufacturers came for access to the oil used in their products; chemical companies came to produce the sulfuric acid used to improve the odor and color of kerosene and expanded to include other products. The dredging of Newtown Creek and construction of railroad facilities attracted still more industries.

The very presence of these companies, as well as their activities, degraded the environment. Some of the harshest pollutants came from every stage of refining petroleum. It is thought that the companies produced three hundred thousand gallons of waste material each week. Before the automobile age, refiners burned off their gasoline. Where it didn't pay to recover waste products, they were discarded on the ground, into the air, and into the waterways. Equally awful, the residue resulting from the improvement of the kerosene was a tarlike substance known as sludge acid, a by-product sold to companies that mixed it with hair, flesh, bones, fish scraps, and other animal parts to produce superphosphate fertilizer. Fertilizer companies dumped remaining wastes into the creek, and at least one of them mixed boiled horse meat with sludge acid and left it to bake in the sun for hours, presumably to enhance decomposition.

Another malodorous and toxic source of pollution was the Laurel Hill Chemical Works established in 1880 on the northern bank of the creek, an enormous refinery and chemical plant that once numbered among the nation's leading copper producers. Built partly on the slag of its own production, it dominated some thirty-six acres with one hundred buildings, where 1,260 employees worked at its peak of production. The sulfur dioxide fumes rising from the sulfuric acid used to refine the copper ore flowed from its massive smokestack and smothered adjacent communities.

Time and again, the public protested. In 1881, *Harper's Weekly* published a series on the vile conditions at Newtown Creek, and, after a period of temporary remission, a *New York Times* article, "The Hunter's Point Nuisances," reported on July 25, 1883, that the "factories were in full blast again, spreading disease and death all around them." Two years later, the *Times* informed local health officials that Standard Oil had begun dumping sludge acid, which covered the banks at low tide. In 1891 a citizen's "Smelling Committee" from Brooklyn took a boat trip on the creek and, from the boat, pinpointed heaps of rotting flesh in doorways, stinking oil tanks surrounded by inky water, other more colorful effluents from factory openings, and troughs in the creek black with industrial wastes. By century's end, Newtown Creek was barren of any life, an ecological wasteland where, as the *Times* article reported, "the discolored and poisonous waters ... now forbid the growth of food for either fish or fowl."

Before the formation of the City of Greater New York in 1898, suburban governments enforced nuisance laws sporadically, if at all. Only the New York City Board of Health pursued the polluters, but it had no authority on Long Island. Attempts to make real progress failed against the powerful corporations. In court cases the Brooklyn Board of Health sided with industry, and nothing much changed until the federal government passed an oil pollution law in 1924 that had been weakened by the protests of industry in its transit through Congress. Regulations notwithstanding, the largest pool of underground oil spills in the country lies today beneath the shores of Newtown Creek in the Greenpoint section of Brooklyn.

The oil came from leaks in the 1940s and 1950s from ExxonMobil and possibly BP Amoco and ChevronTexaco. It covered about fifty-five acres on land and beneath the creek, and it contained 17 million gallons—about 6 million more than the *Exxon Valdez* spill in Alaska in 1989. In 1978, a U.S. Coast Guard helicopter patrol discovered a large plume of oil flowing out of the banks of the Newtown Creek. But others had become aware of the leaks earlier, when gasoline fumes exploded in a sewer on Henry Street in 1950, shooting twenty-five manhole covers into the air and shattering windowpanes in more than five hundred buildings. Since 1981 ExxonMobil has pumped out and sold about 2.5 million gallons of oil while further polluting the creek in the process. The remaining oil sits on top of the Brooklyn aquifer, and while little is known about the health hazards, the deleterious effect on the water quality of Newtown Creek is obvious.

On the north bank of Newtown Creek opposite Greenpoint, the neighborhood of Laurel Hill, now called West Maspeth, has an equally toxic history. The pollution is directly traceable to a century of flourishing industry when it was both convenient and profitable to locate chemical plants and copper smelting facilities in Queens at the generous confluence of flowing water, plentiful cheap labor, and global shipping facilities. It all started after the Civil War, when Dr. William Henry Nichols founded a chemical research and manufacturing operation here. By 1876 he partnered with metallurgist John Brown Herreshoff to develop industrial techniques to break down copper sulfide ore into the component products of sulfuric acid and copper, just when the nation's industrial rise created vast new markets for both. By century's end, success split the company into a chemical corporation that evolved into Allied Chemical, while the metals side became known as the Nichols Copper Company until it merged with its major copper ore supplier, Phelps Dodge, in 1928. During its years of maximum productivity, the Laurel Hill Chemical Works employed about 1,250 employees, but as the industrial age wound down, closure of the plant's smelter in 1964 halved the workforce, and further retrenchments in 1971 reduced the roster to 400 employees. By 1983, Phelps Dodge shut down Laurel Hill, selling the entire parcel to the U.S. Postal Service in 1986.

Alas, the USPS soon discovered dangerously high levels of the heavy metal by-products of copper smelting—lead and cadmium among the most troubling—and the case wound up first in the hands of the New York State Department of Environmental Conservation and then of the U.S. Attorney's Office, who forced Phelps Dodge to refund the purchase, take back its property, and agree to a cleanup, which left the site razed but still polluted, as it remains at present.

MOVING TOWARD A MODERN SOLUTION

Given the city's dedication to commerce, it's not surprising that the first efforts to protect its waters from pollution were meant to promote tourism on Coney Island. Coney was still a true island when the city fathers saw that it was necessary to protect the waters for the health of day-trippers and holiday makers who flocked to the beach hotels and bathing establishments every summer in search of sea breezes and refreshment. The city's first sewage plant was built here in 1887, and another in 1891 at Sheepshead Bay nearby; both used chemical precipitation and were considered models of sewage treatment. In 1903 the first city Sewerage Commission was designated, and by 1920 it had adopted a regional sewage treatment plan based on natural drainage patterns. The Jamaica Bay treatment facility went up in 1927, followed by three more regional plants by 1952. Shortly after a major trunk sewage system opened in Raritan, New Jersey, contemporary reports documented a dramatic increase of marine life in Raritan Bay.

Today, one of New York City's major wastewater treatment plants is located at Newtown Creek. This is the destination of much of the drainage from Manhattan's East Side, which feeds into interceptor pipes leading to the Thirteenth Street Pumping Station, a system that is considered a great engineering feat. At Thirteenth Street and Avenue

D, sewage collected from the East Side up to Seventy-First Street, as well as from Midtown, the Lower East Side, Greenwich Village, and the financial district, is pumped through a tunnel under the East River and up into vats in the Newtown Creek plant, where it is processed. This influx, gathered by 180 miles of sewers, amounts to 170 million gallons a day during dry weather and as much as 300 million gallons under wet conditions. During the 1977 blackout, 828 million gallons of untreated sewage flowed into the East River. And in spite of the fact that New York State ordered the city in 1995 to build a backup generator, another 145 million gallons spilled during the August 2003 blackout. That order for a backup generator has been fulfilled.

The New York City Department of Environmental Protection calculates that New York's 9 million residents discharge 1.5 billion gallons of liquid waste into the sewer system every day. Once dumped, the effluent navigates the six thousand miles of pipes, 135,000 catch basins, and ninety-three pumping stations of the wastewater treatment network toward two possible futures: decontamination at one of fourteen treatment plants or discharge into New York Harbor via one of 494 combined sewer-overflow outfalls. The Department of Environmental Protection's Newtown Creek Sewage Treatment website says that such an overflow event generally happens about once a week and discharges about 500 million gallons of raw sewage directly into New York Harbor each time. To quote the department directly: "When sewage loads exceed the capacity of the Newtown Creek Sewage Treatment Facility[,] trash, pesticides, petroleum products, PCBs, mercury, cadmium, lead, pathogenic microorganisms, and nutrients[,] which reduce the dissolved oxygen content of the water[,] are dumped into Newtown Creek.... Essentially anything that gets washed into the gutters from the street, anything that households and businesses flush down the toilet or dump down the drain, has a fair chance of being expelled directly into Newtown Creek or New York Harbor untreated."

Between the financial considerations and political finagling, the city is still not in full compliance with the Federal Clean Water Act passed in 1972, which calls for secondary treatment to remove 85 percent of all pollutants from incoming sewage. In the interim, New York State's 1992 order for the city to plan, design, and construct a system of facilities to eliminate combined sewer-overflow outfalls by 2013 was stalled by the city's request to postpone that deadline for a fully compliant new Newtown Creek facility until 2022. Upon completion, the site, which treats wastewater from a twenty-five-square-mile area, including parts of Queens and Brooklyn, will expand from thirty-six acres to fifty-three at a reported cost exceeding $2 billion.

Newtown Creek remained in the news as State Attorney General (now Governor) Andrew Cuomo launched legal action against five companies, holding them responsible for environmental damage caused by the half-century-old oil spill there. A notice from the attorney general's office informed ExxonMobil, BP, Chevron, KeySpan, and Phelps Dodge of one corporation's internal study that found the presence in and around Newtown Creek of almost a hundred pollutants, including arsenic, benzene, and lead. Tests conducted in 2005 by the environmental activist group Riverkeeper found toxic fumes rising from the ground above the spill, which led to a separate lawsuit lodged by Greenpoint residents. Local citizens and politicians currently look forward to a thorough cleanup leading to the creation of a recreational waterway.

THINK GLOBALLY, ACT LOCALLY

Important as they are, issues of water and waste hardly exhaust the subject of the serious impact that cities like New York have beyond their narrow confines. The World Trade Center came under attack for the very reason that it stood as an icon of successful international trade and commerce—while symbolizing, as well, the impact of global trading practices, rooted in the wealthier, industrialized nations, on the ecosystems of the world. In no small measure, cities are a major cause of *global* environmental degradation and the resulting ecosystem destruction and species loss.

Yet many examples show that we *can* think locally and set a good example globally. Let us return for a last look at that pier on the Chelsea shore of the Hudson. Perhaps we *could* risk an oyster should one be plucked now from the pilings below our feet. After years of regulators forbidding the harvest of virtually all edible marine life, scientists are struck by the return of an array of life in the Hudson estuary great enough to qualify as the recovery of an ecosystem. Just as the living biomass of Chesapeake Bay is collapsing under a vast, unregulated burden of land runoff loaded with fertilizer and industrial pollutants, the Hudson yields increasing harvests of shad, blue crab, and eel. Why is this? Past decades have seen real progress in cleaning up the river.

Another factor is the regulation of PCBs in the Hudson River. The Hudson is the only major river where no commercial fishing is allowed; for this reason, the eel population boasts a full range of age categories, since the species is not fished. Eels go to the Sargasso Sea to breed and eventually die. The young eels hatch and swim to the Hudson (and other rivers) and live for twenty to twenty-five years. Their bodies become hairy and silvery and grow up to fifteen feet long before they leave to breed for their final time. When stressed, all marine life did plummet in the Hudson: there was a 70 percent drop in the catch between 1950 and 1967. But even when the river was heavily polluted, diversity in the Hudson was little reduced, except for some of the more charismatic animals, like sharks and whales. Now free of excessive human predation in the form of overfishing, marine life is rebounding, and the Hudson River eel population is one of the few worldwide not threatened with severe decline.

CHAPTER SIX

Invasion and Survival

We are in the few remaining dockyards across the Hudson in West New York, the jumping-off point for many of Europe's invading plant and animal species. Colonists have always brought their favorite plants and animals with them, and some other species have simply come along for the ride. Some of the greatest killers of species are simply the invaders that people bring with them, and many of the species surrounding New Yorkers are such invaders. Pigeons, house sparrows, and starlings all came from Europe, as did viper's bugloss, chicory, Queen Anne's lace, and many other weedy plants that grow up each summer along abandoned railway lines and along roadsides. So what looks like a halfway vibrant ecosystem is, to some observant eyes, often not that at all: it is, rather, a motley assortment of ecologically hardy species that early on learned to live a semiparasitic life within human settlements.

Yet life is hardy, and surprising pockets of "wild New York" remain. Famously, Manhattan's tall buildings and bridges are home to roosting peregrine falcons and red-tailed hawks. These species reclaimed old nesting grounds after Rachel Carson's *Silent Spring* started the successful anti-DDT campaign in America. Less dramatic, but far more typical, is the exuberant mix of native and foreign species that bursts forth every spring in open lots, on wasteland, in untended park spaces, and

Figure 64. Pigeons above waiting for a meal below, at Fifty-Sixth Street off Fifth Avenue. Horse-drawn carriages are a major tourist attraction, taking visitors for rides in Central Park. Photograph by Sidney Horenstein.

JOHN TORREY

One of my great pleasures in the distant past was to explore the musty used book shops south of Fourteenth Street. The Strand Book Store today is one of the few reminders (but now not so musty) of that past delight. In one of those stores, I found and purchased (for $2.50) a copy of *A Catalogue of Plants, Growing Spontaneously within Thirty Miles of the City of New-York,* published in 1819. John Torrey (April 15, 1796–March 10, 1873) was the principal compiler of this important early botanical work. As a young boy living in Greenwich (now Greenwich Village), he explored with his brothers the rocks cropping out along Broadway north of Grand Street, looking for garnets, then for plants and insects in the low ground of Lispenard Meadows to the west.

Torrey was a lifetime resident of New York City, where he focused his outstanding career. His interest in botany began early; he visited the Elgin Botanical Garden (located where Rockefeller Center now stands) and studied under Amos Eaton, one of New York State's great naturalists, in the most peculiar way. Young Torrey's father was the fiscal agent of the state prison in Greenwich, where Eaton was in prison for life, convicted (probably falsely) of fraud involving real estate dealings. It was here that young Torrey learned from Eaton about botany and other natural history subjects. With the help of Torrey's father, Eaton was released early and went on to publish many important volumes related to New York State, including a survey of the geology of the Erie Canal (the very feature that helped to make New York City the "Empire City"). Eaton was also the cofounder of and ran the Rensselaer Institute (now the Rensselaer Polytechnic Institute).

While still a twenty-one-year-old student at the Columbia College of Physicians and Surgeons in 1817, Torrey helped establish one of the city's important intellectual organizations, the New York Lyceum (later the New York Academy of Sciences). After getting his medical degree in 1818, Torrey taught chemistry, mineralogy, and geology at West

Point Military Academy (and was an assistant surgeon, a field that did not interest him much). He came to Columbia College to teach chemistry and also taught at the College of New Jersey (now Princeton University). During these years, while the federal government was sponsoring expeditions to unexplored parts of the West and South, almost all the plant material collected in those regions was sent to Torrey to be analyzed and classified.

So thanks to Torrey, New York City became the center of botanical studies in the United States. Upon retirement in 1854, he was appointed the first assayer of the U.S. Mint in New York City. By then, the city's growth was destroying the natural features, flowering fields, and dense woods that had been familiar to Torrey as a boy, so he bought a rural farmhouse and several acres of land on the Palisades above the western bank of the Hudson River. Today, his property is part of Columbia University's Lamont-Doherty Earth Observatory, a major research organization that studies fundamental knowledge about the origin, evolution, and future of the natural world, including cities. Torrey's assay office was located at 30–32 Wall Street in the Branch Bank of New York State, a Greek Revival building erected in 1824. The building's facade was preserved in 1924 in the American Wing of the Metropolitan Museum of Art. When you enter the building, walk up the stairs, and immediately to the left you can look out the window that was part of his office.

In 1867, while Torrey was still alive, the Torrey Botanical Club was formed. And still meets today, as the Torrey Botanical Society, at the New York Botanical Garden. When Torrey died in 1873, he was in his home on the campus of Columbia College at Madison Avenue and Forty-Ninth Street, once part of the Elgin Botanical Garden.   *SH* along even the most degraded waterways. Unlike in Chicago, where vestiges of native prairie can pop up in abandoned industrial sectors, New York's native woodland is less likely to spontaneously reconstitute itself and is far more rarely found. Yet should something happen to this concrete jungle, nature—that is, an entirely reasonable facsimile of primordial nature—stands ready, willing, and able to reclaim the land. Or at least, as we learned from the wildly overgrown, pre-rescue High Line in New York City, discussed in chapter 7, there will be a good mixture of native and introduced plants.

Indeed, throughout the long course of time, nature has seized every opportunity to reclaim and transform Earth's lands over and over again. When the last glaciers retreated, many wild species developed a circumboreal distribution. At that time, a number of the same or very closely allied species occupied both northern North America and northern Eurasia, such as brown bears, wolves, moose, red foxes, red squirrels, three-toed woodpeckers, and more—all a welcome source of familiarity to European settlers in America. When Scandinavian immigrants reached the New World, they too found species that were different but highly reminiscent of those at home; compare the deer and elk, European and North American bison, and the stands of birch and alder so familiar from Scandinavia and typical of the American Midwest.

Looking back at the history of Manhattan, remember that, during the last Ice Age, the Sixth Extinction of native plants and animals started long before the mechanical revolution often blamed for today's crashing biodiversity. The thousand-foot-thick ice sheet that scraped over New York City wiped out all trace of native American earthworm species, opening a niche filled by European species before native species, which had been pushed southward during glaciation, could fully reinfiltrate the north. Yet the evidence is undeniable that wherever *Homo sapiens* treads, species loss and environmental degradation or disruption are inevitable. Certainly in the New World, in both South America and North American—including what is now Manhattan—the mighty mammoths and giant ground sloths disappeared shortly after the arrival of human bands of hunters, whose knife marks on the bones of their prey reveal humanity's early impact upon the balance of nature.

Whether or not there were Native Americans roaming the periphery of the ice fields of New York in search of now-extinct rhinos and mastodons, after the melting glacier left its trailing load of terminal moraine, now called Long Island, the first people to settle the ancestral land of the Algonquin-speaking folk undoubtedly exploited the natural resources of land and sea for their living. The remains of shell mounds at Tottenville in Staten Island and Inwood in Manhattan tell us that these true New Yorkers loved their oysters just as we do today. In hindsight, we assume their predation fell within the realm of "sustainable use," but those first families changed the landscape nonetheless.

By the time the British general Sir William Howe debarked with four thousand redcoats on the East River near Kips Bay in September 1776—on the hunt for George Washington's forces, who were then retreating up the West Side to Harlem Heights—New York already teemed with new species of plants and animals. But most of the change was imminent rather than already evident. Imagine how hard it would be today to miss the fact that just across Central Park there was an army of thirty-five hundred men hard at a forced march? Yet Howe and the American troops under the command of General Israel Putman were unaware of each other moving along parallel routes in the same direction. What stood between them was a heavily wooded low ridge between Columbus and Amsterdam Avenues—the drainage divide of Manhattan Island that can still be seen today along many streets, including Seventy-Seventh, Seventy-Ninth, and Eighty-First Streets. Today the woods are gone, the ridge has been greatly leveled, and Central Park is what remains of that razed wilderness.

Every new boatload of colonial settlers, every fresh advance upon virgin soil, and every inventive application of Yankee ingenuity meant change both for the determined people who came in such numbers and for the land that yielded up its riches to them. To make an unfamiliar land more like home, gardeners planted some of their favorite flowers from the old country; and at a time when home remedies were the only cures available, traditional medicinal plants, too, gained a foothold in the United States. Common lawn weeds came from the Old World, yet are such an accepted aspect of the landscape that they rarely gain mention among specialists as actual invasive species. These include English plantain (*Prunella vulgaris*)—mentioned in Shakespeare as being good for bruises, and commonly called heal-all—and dandelions, which have tonic properties as greens and in flower wine.

A perfect example of how invasive species get established, and of what gives them such a bad name, is *Alliaria petiolata,* which explains itself with its popular names: garlic root, hedge garlic, sauce-alone, and poor man's mustard. Two more names explain its resilience as a weed: Jack-by-the-hedge and Jack-in-the-bush (keep in mind that *Jack* is an old synonym for "knave"). This extremely hardy and widespread Old World native arrived here in the middle of the nineteenth century, a time when culinary traditions were expanding along with the cross-section of southern Europeans. The crushed foliage smells like garlic, while young leaves taste mildly of garlic and mustard and are eaten in salads and pesto. Tender young leaves appear early in spring, so this fresh, pungent green must have been a godsend in the days before refrigeration and mass marketing of produce.

Today, aside from a contingent of foragers and health enthusiasts, people consider it merely a pest—a toxic, virtually ineradicable pest, at that. White-tailed deer reject it, but as they trample it while foraging and disturbing the soil yet more *Alliaria* is sown. Garlic mustard, as a result, has two advantages over its competitors. First, it is favored by neglect while other plants are eaten; and second, wherever *Alliaria* roots, it inhibits growth of the mycorrhizal fungi critical to flourishing native woodlands. Of course, in its own native Eurasian habitats, none of that is true. The mycorrhizal fungi are adapted to the *Alliaria,* and its growth is contained by natural predators not found in America. Having sown the winds of imbalance in nature, we must now reap a whirlwind of garlic mustard fields in America. Research has shown that seeds can live as long as five years in the soil before germinating, so restoration of invaded areas requires long-term intervention. We ourselves are left as the only potential predators capable of extirpating these invaders we have introduced.

Extreme fertility and an ability to outcompete native species explains the great abundance of any number of alien plant species. The hardy wetland perennial purple loosestrife (*Lythrum salicaria*) is virtually impossible to eradicate, since a single mature plant is capable of producing over 2 million seeds in one season, while a tenacious and fast-growing root system allows it to reproduce underground too. Found here by early in the nineteenth century, its presence is variously blamed on ballast dumping by ships in the harbor, on its continuing popularity as an ornamental plant, and on its medicinal uses in treating diarrhea, dysentery, and ulcers. Whatever the facts, by the 1930s, following the construction of waterways, drainage systems, canals, railways, and highways, purple loosestrife spread inland from along the East Coast and now grows in every state except Florida.

Some alien plant species represent a history of failed commercial endeavors. The purple-blossomed princess tree (*Paulownia tomentosa*) was planted not only for its showy flowers but also because of the hope, unfulfilled, that its wood might be exported for carving. A beautiful tree, it is an aggressive grower that resists fire, cutting, and even crushing—the roots and buds of the mother tree readily put out new shoots. *Paulownia* can grow anywhere there's a bit of water, including cracks in rocks and steep cliffs. The ubiquitous, brambly, multiflower rose (*Multiflora rosa*) springing up all around the metropolis stands as witness to an era when its spiky, impenetrable canes fenced in the cattle and horses once corralled within the city limits. Messy white mulberry trees (*Morus alba*), which drop their pallid, insipid fruits on city sidewalks in abundance every spring, were planted in the colonies during the reign of James I, around 1619, in the first of several short-lived efforts to establish a silk industry here. At least the silkworm did not live on to become a pest: after more than five thousand years of domestication, the moth has lost its ability to fly, and silkworms are extinct in the wild.

Figure 65. Ginkgo tree (*Ginkgo biloba*). Originally from China, this species is considered a living fossil, a phrase originated by Charles Darwin to describe plants and animals that had a wide range in the geologic past but are now limited, and which have not changed much from their ancestral forms. This beautiful specimen, probably planted in the 1860s, stands in Isham Park in northern Manhattan, where it can be seen overhanging Broadway just north of Isham Street. In fall the leaves' color is a striking golden yellow. Photograph by Sidney Horenstein.

Interestingly, the maidenhair tree (*Ginkgo biloba*), native to China, whose fruitlike seeds are equally messy and furthermore stink when ripe, never shows up on invasive species lists. Asian immigrants to New York City have been seen fiercely arguing over the harvest of nutrient-rich ginkgo nuts fallen to the street, and city fathers entertain an endless stream of seasonal complaints from residents every fall about the fetid odor of ginkgo seeds. But efforts are now made to plant only male specimens on New York City streets. The ginkgo, a beautiful, long-lived tree with uniquely fan-shaped foliage and great resistance to pollution and infestation, is destined to remain a mostly welcome fixture.

Betty Smith's 1943 novel, *A Tree Grows in Brooklyn,* compares her feisty young heroine with the unconquerable tree-of-heaven, *Ailanthus altissima.* First planted in North America in 1784 by William Hamilton on his Philadelphia estate (currently the Woodlands Cemetery), it arrived from China via England and was introduced to New York in 1820. *Ailanthus* is the ultimate opportunistic hardy grower whose flourishing habit has made it a popular choice to plant in cities. The tree breeds clonally, rapidly producing dense thickets that can take over virtually any landscape. Poor soil is not a deterrent, and it is not uncommon to find trees sprouting from sidewalk cracks throughout Manhattan and the outer boroughs. Once established, *Ailanthus* discourages other species by producing growth-inhibiting toxins. This aptitude for outsmarting native species is a common complaint against aliens. Spreading Russian olive trees (*Elaeagnus angustifolia*), for example, overwhelm natives, interfere with their nutrient cycle, and soak up the available water. Birds readily eat their fruits, but naturalists find that avian diversity is greater where native plants remain.

Similarly, the common, glossy buckthorn (*Rhamnus frangula*), damned as "the demon weed tree" by one volunteer eradication group, is prolific in producing berries that birds eat and disseminate throughout the Northeast and Midwest. Buckthorns, too, crowd out native species, producing dense shade that effectively prevents plant succession and reforestation. The same is true for that creeping menace kudzu (*Pueraria lobata*), a prolific vine of the bean family. First promoted at the 1876 Philadelphia Centennial Exposition as an ideal fodder and ornamental, this life-smothering climber gained a wide foothold in the southern United States and provided temporary livelihoods to workers in the Civilian Conservation Corps, who planted kudzu there against soil erosion under a work relief program set up by the FDR administration. To date, the numerous kudzu sightings in metropolitan New York have been considered harmless in the face of winter dieback, which eliminates seed viability. But given the continued warming trend, the plant remains a threat for the future.

With a passion for novelty and only the best of intentions, generation after generation of American gardeners has introduced fascinating and beautiful new plants that escape to wreak unforeseen consequences upon our native soil. The handsome big white flowers and spreading growth of the horse chestnut (*Aesculus hippocastanum*) make it a popular ornamental, but its weak limbs, large leaves, and toxic nuts litter the ground; its dominating size robs native plants of needed light, water, and nutrients; and anyone tempted to drink a tea of its leaves, sprouts, or nutmeats will be poisoned—possibly fatally. Diseased horse chestnut leaves also tend to develop unsightly brown blotches in the summer. Yet, a testament to earlier horticultural standards of fitness and beauty, Central Park is full of horse chestnuts happily blooming in spring and casting their noxious nuts to the ground every fall.

Nor is it hard to find other tree species, in Central Park and throughout the city, that have escaped from unwary gardeners. The Bradford, or Callery, pear tree (*Pyrus calleryana*), snowy with spring petals and gracefully tapering, was so widely planted that it managed to cross-pollinate with fertile pears and has escaped to compete with other trees. Birds love the berries of Japanese barberry (*Berberis thunbergii*) and oriental bittersweet (*Celastrus orbiculatus*), found in countless urban and suburban yards. They have spread the barberry, with its choking shade, and the bittersweet, with its strangling vines, far and wide. Easily grown survivors such as porcelainberry (*Ampelopsis brevipedunculata*), mugwort

(*Artemisia vulgaris*), and Asiatic dayflower (*Commelina communis*) are not only capable of pushing through any crack in the city's sidewalks, but they also take over the city's meadows, since the native species that do remain cannot compete.

When Fort Tryon Park underwent a grand makeover in 1988, it was overgrown with Norway maple trees that entirely blocked the magnificent view across the Hudson just north of the George Washington Bridge. This sixty-seven-acre high point in Manhattan was surely wooded when Revolutionary forces fought the Battle of Fort Washington there in 1776, and it was subsequently named by the victors for the British major general Sir William Tryon, last governor of colonial New York. The nineteenth century saw the taming of this landscape with the construction of a grand mansion, much stonework, and manicured grounds. After the house burned down in 1926, John D. Rockefeller, who owned the land, hired the Olmsted brothers to create a park, and established the Cloisters Museum at the park's northern end. To preserve the splendid view, he even bought land across the Hudson River, which was added to the already existing Palisades Interstate Park. Sadly, by 1983, when park restoration began under the aegis of the Greenacre

---

FORT TRYON PARK

Manhattan is endowed with many great parks, large and small, each with some compelling aspect, whether geological, historical, or political. One of my favorites is Fort Tryon Park, located in Manhattan a little more than a half mile north of the George Washington Bridge. It's a relatively small park, but beautifully designed to include wide and spacious parkland walks, and paths narrow and rustic, situated on a northern schist ridge, where it captures the best of its setting.

The plateau on which the park rests has rocky, steep slopes on three sides, and from the top and sides the views are stunning. To the west is the majestic Hudson Valley, the southernmost glacial fiord in North America, and a backdrop of the Palisades cliffs in New

Jersey. Looking northward, one sees a drop-off in topography, caused by the Dyckman Street fault. Beyond that, the ridge regains its height where it becomes Inwood Hill Park, which contains elements of the last remaining forest on Manhattan Island. To the east are the Inwood Flats, underlain by easily eroded Inwood marble, where you can catch a glimpse of the northern edge of Fort George Hill, as well as a view of the valley of upper Broadway and, in the distance, the varied topography of the Bronx. If, on a clear day, you look eastward into the far distance, you can see a faint hill marking the location of the southernmost tip of the Ice Age glacier that entered our area 21,500 years ago.

The park was given to the people of New York City by John D. Rockefeller Jr., who admired this area and wanted to preserve it as a park. Recently, David Rockefeller made a generous gift to help maintain the site. The park was partly created to house the medieval branch of the Metropolitan Museum of Art, the Cloisters, a unique museum in an outstanding setting. Fort Tryon Park was designed by Frederick Olmsted Jr., a well-known landscape architect in his own right, although the plan was governed by the concepts of park design established by his father, Frederick Law Olmsted, in Central Park. The four design elements found in this naturalistic park are the beautiful (small open lawns), the picturesque (wooded slopes), the sublime (vistas of the Hudson Valley and the Palisades), and the gardenesque (the Heather Garden). *SH*

Foundation and the Parks Department, with the help of many volunteers, the highly invasive Norway maple and sycamore maple trees (*Acer platanoides* and *Acer pseudoplatanus*) had mostly taken the place of what would have been a forest composed of such trees as sassafras, hickory, red and black oaks, American beech, tulip poplar, and the native red maple. Alas, while Norway and sycamore maples may be loved for their flaming fall foliage, they overpower our native trees.

IF YOU CAN'T BEAT IT, EAT IT

A New York character by the name of "Wildman" Steve Brill has been specializing in the edible weeds of Central Park and the New York City environs for the past quarter century. Today, he still gives guided weed-gathering tours, complete with recipes for preparing your gleanings. But, as Steve writes in *Raw Foods News Magazine,* he had to earn the privilege: "My big break came at 4 PM, March 29, 1986, when two undercover park rangers who had infiltrated a Central Park tour arrested and handcuffed me for eating a dandelion. The police fingerprinted me and charged me with criminal mischief for removing vegetation from the park, but I had eaten all the evidence, so they released me with a desk appearance ticket pending trial."

These days, the Wildman has a website, several published books on foraging for wild spring greens, and guides to finding just about anything edible growing in and around New York City. It's amazing what you can eat from the sidewalk or vacant lot! Brill's free online app Wild-EdiblesLt features fourteen common weeds worthy of gathering for consumption as field greens or potherbs. Only two of them are native plants; all the rest came from Europe or Asia. The fact that the word *epazote* (*Dysphania ambrosioides*) derives from the Native American Nahuatl word *epazōtl* indicates the Mexican origins of this potherb commonly cooked with bean dishes. However, its frequent appearance in Central Park and Prospect Park probably dates only to the 1970s, in tandem with an influx of Central American immigrants. Imports such as wild garlic (*Allium vineale*) and poor man's pepper (*Lepidium virginicum*) may originally have been cultivated for the table, but others, such as chickweed (*Stellaria media*), curly dock (*Rumex crispus*), and shepherd's purse (*Capsella bursa-pastoris*) undoubtedly first hitched a ride here, whether in a ship's ballast or bags of grain or stuck to the bottom of birds' muddy feet. Whatever the agent of transport, a certain sense of inevitability must be acknowledged, since all are widely distributed around the globe today, and shepherd's purse is deemed the second-most common weed on Earth.

The plants Brill mentions that have medicinal uses, including the common plantain (*Plantago major*) and sheep sorrel (*Rumex acetosella*), are documented as part of the pharmacopoeia of the colonial era. In the case of plantain, in fact, the endemic blackseed plantain (*Plantago rugelii*), which was both eaten and used as a tonic by Native Americans, was supplanted by the broader-leaved common European plantain by 1700. Other edible plants on his list arrived as ornamentals, such as the tall orange daylily (*Hemerocallis fulva*) and the less conspicuous blue Asiatic dayflower (*Commelina communis*), and then later escaped cultivation. The very worst offender in the once-ornamental category is Japanese knotweed (*Polygonum cuspidatum*), which behaves much like bamboo—it's nearly impossible to eradicate, wildly invasive, able to pierce concrete, and grows as much as nine feet high. Unlike bamboo, the plant does die back in winter, but it comes roaring back every spring, spreading farther and stronger than ever. Although the shoots are edible, they are very sour, and even Brill admits that a little goes a long way.

BLIGHTS, MITES, RATS, AND BAD BUGS

Two beloved native American shade trees, the elm and the chestnut, have given street names to half the towns and cities across the nation, including a Chestnut Street in both Brooklyn and the Bronx and a former Elm Street in Manhattan, now renamed Lafayette Street. Before 1904, when the chestnut blight fungus (*Cryphonectria parasitica*) was first discovered infecting a tree in Central Park, some 3.5 to 4 million American chestnuts graced the forests, parks, and streets of the nation. By 1940, a lofty tree that could easily reach a height of fifty feet below its branching crown—a tree whose wood was prized for its beauty and workability, a tree preferred for building barns, and whose single trunk could yield an entire boxcar full of planks—was effectively extinct. The killer fungus spreads easily on the wind and remains dormant in the soil to this day. The Japanese chestnut, which resists the fungus, is both the probable agent of infection and the potential genetic savior. New crosses between the Japanese and American chestnut were recently planted in Central Park and seem to be doing well.

Dutch elm disease, another fungal infection that originated in Asia, was noticed killing European elms sometime after World War I. This blight is "Dutch" only because it was identified by a Netherlands scientist in 1921, and because the original source of infection may have been a shipment of European elm logs from Holland in 1928. Since the discovery of the disease in Cleveland in 1930, estimates of elm destruction on the American continent have reached as high as 77 million. Actual agents of the spread of the fungus are mainly native and alien elm bark beetles and the transport of infected firewood or the use of infected pruning tools on the part of commercial arborists. Amazingly, the magnificent grand allée of towering urn-shaped elms along the Mall at the heart of Central Park remains intact to this day, along with hundreds more elms scattered in groves and singly throughout the park—nearly seventeen hundred in all. Kudos to the park keepers for their care and vigilance in beating back the constant threat of infection, but perhaps we can also thank the concrete jungle outside the park's borders. The infection spreads not only through the short flight of beetles from tree to tree but also from root to root, since congregated elms literally grow together underground. Thus entire forests of American elms (*Ulmus americana*) have died, while most of those isolated in the very midst of Manhattan bow only to old age and shattering storms, like the storm in 2009 that split an elm tree in two.

Ash trees, including green, blue, purple, and white species, account for about 7 percent of the trees in New York State, and they are all prime targets of a beautifully iridescent but deadly beetle from Asia, the emerald ash borer (*Agrilus planipennis*), which measures a third of an inch long. Destruction was well under way, and reports were filtering in from the field, when the culprit was identified in Michigan in 2002. Since then, the U.S. government has estimated that the nation has lost more than 50 million ash trees. How the beetle got here is anybody's guess; the New York State Department of Environmental Conservation

Figure 66. American elms near Central Park's Mall. Located between Sixty-Sixth and Seventy-Second Streets, the Mall has one of the largest elm plantations in the United States. Widely planted as a shade tree, one with a striking silhouette, this species was decimated by introduced Dutch elm disease, which began to rapidly take its toll throughout the species's range in the 1930s. Photograph by Sidney Horenstein.

speculates it may have come from Asia on infested ash wood used in shipping pallets. But now that it's here, the beetle mostly travels in firewood that people move around—in spite of a ban on transporting firewood in many of the afflicted states.

The USDA took a "bioterrorist" approach to the emerald ash borer in eight states in 2013, releasing more than 150,000 stingless wasps bred in Michigan from wasps imported from the forests in China where the ash borer occurs naturally. The wasps have been feasting solely on ash borer eggs and larvae ever since. Meanwhile, in New York City, the Central Park Conservancy's director of operations made the executive decision by 2006 not to plant more ash trees in the park, even before the advent of the ash borer, because global warming was already truncating the trees' potential 250-year life span. However, other park agencies around New York City continue to plant ash trees.

Another pallet-invader from China, the Asian longhorned beetle (*Anoplophora glabripennis*), was first discovered in 1996 in Brooklyn, where it is now widespread. Extermination can only be achieved through chipping or burning the infected wood, and a number of affected trees have already been removed from Central Park and other Manhattan locations. If the ash tree crisis ultimately robs us of ashen baseball bats and furniture, the even more vital New York State maple syrup industry will suffer too, since maple trees succumb to the Asian longhorned beetle, as do the horse chestnuts, birches, poplars, willows, and London plane trees the beetle prefers.

Some bugs, specifically bedbugs, are more annoying than deadly—but only because people don't actually die of bedbug bites! In fact, the greatest risk with these bugs is pesticide poisoning in those who over-react to an infestation. Once bitten, the worst that can happen to victims is superficial infection from scratching the bites, but try telling that to a New York City apartment dweller who has been visited by *Cimex lectularius* and see where it gets you. In fact, just come to New York City and you may encounter the horror yourself. A 2010 article in *The Week* magazine listed a dozen high-profile locations where bedbugs had been found, including Lincoln Center, the United Nations, the Empire State Building, the New York Public Library, and the famed shopping destination Bloomingdale's. And in a nice apolitical twist, *Cimex* has been found colonizing the television studios of both CNN and Fox. Since bedbugs willingly crawl from one carrier to another, no movie theater, subway train, or taxicab is entirely safe, and the insects have been reported in many city hotels, luxurious and otherwise.

Bedbugs have probably been around nearly as long as humankind has, but they disappeared in the United States before World War II, when vacuum cleaners became widely available and DDT was widely in use. The worldwide *Cimex* resurgence in temperate climates is unexplained but may relate to reductions in DDT use and increased urban crowding. Certainly, the massive amount of postwar international travel and trade is a major source of reintroduction in New York City. Visitors to the city may beware, but residents may be comforted by the knowledge that the cockroaches found in virtually every apartment house in the city delight in feeding upon bedbugs.

As for the cockroach (*Blatella germanica*), it has been an inevitable feature of any urban setting since time immemorial. Indeed, this eternally hardy arthropod long antedates humanity and all our nearest relatives, since cockroaches are true living fossils whose origins date as far back as the Paleozoic era. Given the extensive travel known in ancient days, it is entirely possible that America's earliest Viking visitors brought along seafaring "water bugs," who may already have hitchhiked their way to Europe from a probable Southeast Asian homeland. Also called black beetles, brown roaches, German cockroaches, and Croton bugs, the common roach's alternate, New York–based name is a reminder of the many laboring immigrants from Europe who helped build New York City's first upstate water reservoir, the Croton Dam. Of course, these shipborne settlers could not have avoided bringing ashore a cargo of undocumented *Blatella germanica.*

Rats, too, are perennial stowaways who seem never more comfortable than when in the proximity of their fellow mammals—humankind. Most likely originating somewhere in Asia, New York City's familiar Norway rat (*Rattus norvegicus*) has been documented as present in the New England colonies by 1775, and its remains have been traced back to third-century London. While the invading rat's environmental impact on a pristine environment such as the Galápagos Islands, where depredations upon the eggs of birds and of other small animals is incalculable, any effect of these commensal animals' habitation on us in New York shrinks to insignificance compared to our city fathers' paving over nearly the entire island of Manhattan.

Even merely stinky bugs make unwelcome home intruders here. Case in point: the brown marmorated stink bug (*Halyomorpha halys*), a half-inch-long, winged beetle from Asia first spotted in Allentown,

Pennsylvania, about a decade ago. While these bugs are also eating crops in the Midwest, the main buzz on them results from what can at times amount to hundreds of beetles attempting to seek domestic shelter in cool weather. As the name implies, squashing these little invaders produces a pungent, disagreeable odor.

When the first Dutch farmers broke ground in Nieuw Haarlem in 1658, they did not neglect the craft of beekeeping, since pollination was vital in the orchard, and honey was always welcome in the kitchen. By the year 1899, in a Victorian nod to commerce, tradition, and healthful pastoral pursuits, Central Park had a beehive on display, handsomely mounted in a chalet-style display case. Yet in 1999, then-Mayor Rudy Giuliani put the honeybee (*Apis mellifera*) on a list of wild animals banned from New York City; this was a list that included alligators, lions, and grizzly bears. Naturally, as many as 650 apiarists (beekeepers) kept bees anyway. Still, it was a source of joy to beekeepers around the city when the decadelong ordinance against raising bees in the metropolis was lifted in 2010.

Aside from the risk of bee stings, now considered minor by the Department of Health, it's tougher to raise these busy, self-supporting insects than you might imagine. A host of potential bee ailments exists, and the cumulative effects can lead to colony collapse. The latest threat is honeybee mites (*Acarapis woodi* and *Varroa jacobsoni*), which came from the Old World too recently for the bees to have resistance to them. None of this is surprising, considering that no native American honeybee exists. All those here are descended from European imports, which arrived mainly in the nineteenth century. Our so-called wild bees are actually feral, or escaped, immigrants. As America's westward expansion progressed to the Rocky Mountains, settlers conveniently found already established swarms of honeybees, escapees from more easterly farms, ready to pollinate their equally imported crops and fruit trees. Plummeting bee populations continue to worry commercial agriculture and apiarists worldwide. And in New York City, while the loss of income for a modest harvest of local honey at around ten dollars a pint may not be critical, think of the many plants growing in New York City that have always counted on honeybee pollination to thrive—and what their future may be.

WHAT'S IN THE WATER?

Famously, Manhattan is an island. The surrounding waters make that so, yet the Hudson River, the East River, the wetlands of Jamaica Bay, and other waterways found within the city limits also exist as environments unto themselves. Today we revere our few remaining metropolitan wetlands as nature's filters and prize them as recreational retreats. This very modern viewpoint reverses some three centuries of fear, loathing, and land grabs. "Unimproved" swampy land has traditionally been seen as dangerous and pestilential, infested with mosquitoes, tricky to navigate, a shelter for runaway slaves or renegades, and impossible to cultivate. Land reclamation started in Manhattan as soon as Peter Stuyvesant became the colony's governor, in 1646, and mandated construction of a canal along today's Broad Street. The excavated debris extended the shoreline, part of an ongoing process that, most recently, as we have seen, yielded Battery Park City, built in the 1970s on the bedrock removed in the construction of the original World Trade Center and other buildings in the area, as well as on sand dredged from the Rockaway inlet. By the time of the American Revolution, the city's population of some thirty thousand New Yorkers had placed a premium on real estate, and well-heeled residents seized the offer to buy city "water lots" in return for filling in the shoreline and erecting buildings.

Calculations vary, but it's likely that today no more than 1 percent of New York City's historic freshwater wetlands and only about 10 percent of its original tidal wetlands remain. The greatest bulk of surviving city marshland, in and around Jamaica Bay in Queens, has been protected since the Gateway National Recreation Area was created in 1972, but rising sea levels and erosion continue to claim about 20 acres of the recreation area every year. It has been estimated that the combined crea-

THE BATTERY

I enjoy the same vistas that George Washington did, strolling down from Cherry Street to the Battery to take in the fresh breezes and view the fleets of boats and ships plying the waters of Upper Bay at the southern tip of Manhattan Island. Now, of course, there are fewer vessels. The most common are the Staten Island ferries and the NY Waterway boats. I've always wondered what it was like in the bay when it teemed with fish, sharks, harbor seals, probably whales, and the birds that made a living there. By all accounts, the large numbers of birds and other species must have been the waterside visitor's constant companions and, during the migration months, a wondrous sight.

The views of Upper Bay from the park are outstanding, and they offer a perspective on all the major geological features of metropolitan New York. Although no rock exposures remain in the park, there is schist below my feet, while the topographical features indicate what the major geological features are: the terminal moraine of the glaciers in Brooklyn, the distant Palisades in New Jersey, and the resistant Staten Island serpentinite underpinning the highest parts of Staten Island. Among several places nearby where one may view these features, my favorite is the roof deck of the Wagner Park Pavilion in the adjacent Battery Park City. From here, you can see the Verrazano-Narrows Bridge crossing from Brooklyn to Staten Island. A relatively narrow constriction separates Upper and Lower Bay and divides Brooklyn from Staten Island. The resulting strait, the Narrows, gives the bays and port an hourglass shape. When the bridge was built, it was the longest suspension bridge in the world, and the towers had to take into account the curvature of the earth—making them slightly farther apart at the top than the bottom.

The port of New York (and New Jersey) was created by people, but the harbor was made by geology. One of the outstanding features of Battery Park is the fortification now known as Castle Clinton, built during the War of 1812. Fortunately, this historic structure has been preserved, and it now serves as the ticket booth for tours to the Statue of Liberty and Ellis Island. In one seldom-visited side room of the castle, a series of three-dimensional models shows the evolution of Battery Park and the adjacent financial district. Virtually all of Battery Park is landfill.

When George Washington arrived in the city in 1789 to take the oath of office, New York City was the capital of the United States—but only for a little more than a year. Washington was fifty-seven years old, still energetic and physically active, and he exercised regularly. He walked through the streets of the city, and his diary notes many occasions when he walked along the Battery—along what today would be, in part, State Street. It was only six years after the Revolutionary War had ended, and New York was not an entirely pleasant place to be. In fact, it was difficult to find a suitable place in the city for Washington to live and conduct state business. The city was in continuous upheaval, undergoing expansion and change. Hills were being leveled, valleys filled in, sidewalks widened in places where possible, and additional streets created by landfill along the waterfront. Slaves carried pots of waste on their heads to be dumped into the river, and tea-water men still supplied potable water by the barrel. It was a shabby little town of about thirty thousand, with narrow and crooked streets that were poorly paved and sparsely lit. The houses were arranged in ragged rows, the wharves were unclean, and pigs roamed the streets eating garbage thrown from the houses.

Yet Washington was fond of the city. He was perhaps even fonder of the all-day carriage rides through the rural countryside with Martha and the grandchildren, which he called the "fourteen miles round." These took him up to about where Grant's Tomb is today and back along the west side, letting him view the farms, speak to the farmers, and enjoy the gardens and dense woods. In a year's time, the capital was moved to Philadelphia, and Washington never returned.   *SH*

THE SEAWALL

For many years I have narrated educational outreach programs on a chartered Circle Line boat in New York's harbor for members of the American Museum of Natural History. Although the emphasis is on geology and the environment, it often includes historical associations. On these voyages, I like to say that geology made the harbor, but people made the port. As we leave Pier 83 at the foot of Forty-Second Street, I often point out the tidal marks on the seawall (the bulkhead), as well as on the pilings, explaining that these marks reveal the position and condition of the tide, which affects the timing of the cruise. I also indicate, to the surprise of many of the participants, that they are looking at one of the great engineering works in and directly adjacent to Manhattan Island. And that they had actually stood on this modern granite seawall while waiting to board the vessel—a wall that was part of a plan introduced in 1871 to rebuild the port facilities, which in previous centuries had been more or less unregulated.

To modernize the port for the future (and accommodate the ever-increasing size of vessels) and maintain its preeminent position as the number one harbor in the country, the city had to reconfigure the port's facilities: the pier lines and bulkhead lines, the piers and pier spacing, the docks and wharves, and the onshore facilities (including constructing wider access roads). The first bulkhead had been built by the Dutch in 1654. By 1870, the condition of the port had so deteriorated that the piers and other facilities became public hazards; rats abounded, and the tumbledown structures that stood in an intricate network became the lairs of river thieves and pirates.

Many groups called for a complete renovation of the port, starting with reforming the great number of city agencies involved, which had made it nearly impossible to plan and administer the waterfront. In 1870, the legislature not only created the Department of Docks but also reshaped other city agencies into the Department of Parks, Department of Public Works, and Department of Buildings.

The Department of Docks was given the job of creating a master plan for the entire waterfront; this required a person with leadership abilities and engineering skills to administer the monumental task.

General George Brinton McClellan was appointed. He had been general in chief during the Civil War for two years, a designer of fortifications, a surveyor of river and railroad routes, and the 1864 Democratic presidential candidate, and he had also worked on developing a new type of steam-powered warship. Given a mandate to correct the present conditions of the port and create a new system of facilities, he opened public meetings and called for and received over seventy very different plans. In the end, the plan he executed was his own. It consisted of creating a new edge along the shoreline using landfill, upon which would be built wide river streets, a masonry river-wall that would encircle the island, and wooden piers at given intervals.

McClellan was on the job for only two years, but his plan was carried out by his successors—with modifications as work proceeded and as new technologies were introduced. By 1878, the department had surveyed 13,700 feet of the waterfront on the Hudson, East, and Harlem Rivers, and it had established new pier and bulkhead lines after determining the water-current and tidal conditions in the harbor—in fact, all of the information required for construction. By 1900, seventy-four miles of waterfront had been surveyed. Hundreds of cores of river sediments had been retrieved so that builders could better understand the depth of the river's bedrock and what dredging and removal of natural obstructions would be required.

A major aspect of the original plan was the construction of the monumental and continuous masonry bulkhead. Although the plan specified concrete, some of the wall was made of granite blocks. The original plan also called for a twenty-eight-and-a-half-mile seawall from the Battery to Sixty-First Street along the Hudson, and to Fifty-First Street along the East River. The Department of Docks achieved its original goal just as World War I began, in time to make New York City a principal port of embarkation for troops and equipment on their way to Europe.

Not all of the new port facilities were extended on landfill, however. To accommodate longer piers in the Chelsea area, the old bulkhead was removed, the shoreline carved out, and a new bulkhead created on the indented shoreline, a process begun in 1902 and completed in 1910. The curving shoreline to the north of the new Chelsea facilities is still obvious today. To make the changes to the present bulkhead position, approval was needed from both the state assembly and the federal government. This complication occurred in 1888, when the federal Rivers and Harbor Act gave the Army Corps of Engineers jurisdiction over all activity involving the country's waterways. The Corps alone had the power to determine bulkhead and pier lines for harbors. As a result of the enlarged piers, Chelsea became a passenger ship terminal in the early 1900s. The *Lusitania* left port here before it was torpedoed by the Germans. This was also the destination of the *Titanic*. When it sank, its survivors were picked up by the *Carpathia* and brought to Chelsea.

The seawall gives rigidity to the island's edge but also protects areas inland from storms—although not always. On the museum cruises, I often speak about climate change: one effect upon the New York area is that sea level has risen about one foot since 1900. Although New York City is not subject to the kind of massive movements of water created by earthquakes and the resulting tsunamis, such as the one that devastated Japan in 2011, it is nevertheless prone to flooding by storm surges. New York City's seawalls are only five feet above mean sea level, and storm waves have overtopped them, especially in southern Manhattan—flooding streets and subways during, for example, the nor'easter on December 2, 1992, as well as during the devastating Hurricane Sandy on October 29, 2012. The highest storm surge on record occurred during the September 3, 1821, hurricane, when water rose thirteen feet at the Battery in

> just one hour and flooded lower Manhattan as far north as Canal Street. I have never liked ending a cruise on a down note, but if warming and the resulting rise in sea level continue, the current bulkhead, at least at the southern end of Manhattan (and elsewhere), will not be adequate. *SH*

tion of landfills, building construction, and airport runways ate up about 1,174 acres of tidal Jamaica Bay land between 1900 and 1974. Moving to terra firma, federal figures indicate that an original endowment of some 224,000 acres of freshwater bodies in New York City has fallen to a mere 2,000 acres or so, thanks to the extensive filling, draining, and hardening of shorelines associated with countless development projects related to housing, commerce, industry, and transportation works.

Given the paucity of landlocked freshwater habitats remaining—these are mainly a few remnant Staten Island outliers—little of the original assemblage of freshwater plants and animals remains. However, the salty and brackish waters of the Hudson-Raritan Estuary, the north and south shores of Long Island, and the East River flanking Manhattan are another matter. As we approach a half century of remission in the pollution of habitats, thanks to pollution control plus cleanups like dredging and waste removal, and to development restrictions and further environmental controls, the promising returns of aquatic flora and fauna remind us of both what is possible and how much is at risk.

Take the extreme example of Gowanus Canal in Brooklyn, once a winding creek that emptied into a tidal inlet fringed by meadows and marshlands overlooking the Upper New York Bay, generally known as New York Harbor—in many ways a story paralleling that of Newtown Creek, encountered in chapter 5. Some of the Dutch colony's first international commerce came from here: Gowanus village fishermen harvested the large and bounteous oysters for export to Europe. As

Brooklyn expanded, by mid-nineteenth century, to the nation's third-largest city, it gained great commercial advantage with a shipping lane dug along the creek from the harbor and 1.8 miles into the city center. Demand was also keen for shoreline building lots created with detritus from the Gowanus Canal.

Soon the banks of the canal filled with commercial enterprises whose wastes and effluents ranged from dirty to deadly. Culprits included paint and ink factories, coal yards, tanneries, and various chemical plants, including the first U.S. producer of chemical fertilizers. In theory, a mechanical system helped sweep away the mucky canal water by pumping in water from the East River. But critics were commenting by 1910 that the water was nearly solid with waste, and so the brick-lined Gowanus flushing tunnel and pump were installed in 1911 to flush out the filth. Still, the pollution always overwhelmed the solution, and the pumping system broke down entirely when a city worker reportedly dropped a manhole cover on the pump sometime in the 1960s. Since then, sporadic efforts to address the raw sewage that overflows into the canal from fourteen sewer drains during heavy weather have finally come to fruition as we write.

In 2010, Gowanus Canal gained Superfund status in spite of New York City's long-standing empty protests that it could do the job better and quicker. Nonetheless, following the passage of the Clean Water Act of 1975, and the public's increasing sensitivity to the environment, this much-abused waterway has attracted a growing number of visitors—artists, hikers, boaters, anglers, and nature lovers. While the city government still warns against any exposure to waters contaminated by such pathogens as *E. coli* and gonorrhea, the waters once again harbor oysters, white perch, herring, striped bass, crabs, jellyfish, and anchovies. In 2007, there was even a minke whale who beached itself and died in the canal. Needless to say, nothing caught in the Gowanus Canal is safe to eat.

The Chinese mitten crab (*Eriocheir sinensis*) is a water-intake threat with a growing presence in the Hudson River in the New York region, first dredged up in 2005 by a Chesapeake Bay crabber. Mature specimens develop hairy growths on their claws, which accounts for the "mitten" moniker. Native to the Pacific coast of China and Korea, the crabs probably arrived in ballast water. They live in freshwater for the first two to five years of life and migrate to salt or brackish water to breed. The crabs can walk some distance on land and have been found as far as two hundred miles away from salt water. Their aggressive nature makes them a potential threat to the native blue crab population, although research on the mitten crab's full impact is still inconclusive. Meanwhile, the New York State fish and wildlife regulations prohibit "possession, importation, purchase or sale or offer of purchase or sale of Chinese mitten crab whether dead or live," while federal law prohibits their interstate transportation.

But there are even more menacing crabs here: the European green crab (*Carcinus maenas*) is considered one of the world's ten worst invasive species. According to the U.S. Department of Agriculture, it was discovered on the Atlantic coast of America in 1817. By the 1950s, the green crab's predations were causing a notable effect on shellfish harvests, particularly clams, scallops, and young oysters. In Massachusetts, the ongoing depletion of soft-shell clams has been linked to an unwelcome abundance of green crabs. They are present in the East River in New York, as well as in coastal streams along the lower Hudson. Furthermore, the little silver-dollar-sized Asian shore crab (*Hemigrapsus sanguineus*) occupies the same territory, produces as many as four clutches annually of up to fifty thousand eggs, consumes a similarly broad diet, and enjoys the absence of native parasites that would plague it. In short, with twice the breeding season of native crabs and broader feeding and habitat parameters, these foreign crabs outcompete our native species and continue to contribute to their decline.

It's the same story with invasive water plants: they arrived here from foreign lands, they choke out native plants and alter habitats, and they interfere with human commerce and recreation. Granted, the greatest damage to New York City's aquatic plant habitats is a result of economic development over the centuries, but in some cases that degradation only broadens the opening for invasive species. Such is the case with the lovely but pernicious purple loosestrife (*Lythrum salicaria*), encountered earlier in this chapter.

The case of the common reed (*Phragmites australis*) is a bit more complex. In fact, *Phragmites australis* is now a global plant, but recent classification studies reveal that there is a specific native subspecies dating back to prehistoric times that is slightly less robust than the more recent invader. So it's hard to say whether our local common reed diminished because of harsh environmental conditions or because the invader is to blame. Whatever the history, the fact remains that the reeds are taking over. All too often, where once there was sandy beach or unclogged estuarine habitat, now there are acres of *Phragmites*. Once again, nature's diversity is the loser. From Central Park to the shores of the Hudson and East Rivers, those plumed, grassy-looking, head-high stalks take the place of what was once a diverse assemblage of grasses, cattails, reeds, and shrubs feeding and sheltering now-long-gone aquatic and avian species.

Looking into the water, we find still more menaces afloat. One of the nastiest is the vigorous water chestnut (*Trapa natans*), which was first documented here in 1879 by the noted Harvard botanist Asa Gray after a worker at the Cambridge University Botanic Garden planted this Eurasian import in neighboring ponds and other local waterways. The rest, as they say, is history. If you've ever gone to a New York City beach such as Coney Island or Jones Beach and painfully stepped on a four-pronged, chestnut-sized, hard-shelled sort of nut, that's the water chestnut, or caltrop. The plant is found throughout the region floating on the surface of fresh or brackish water in broad mats of verdant diamond-shaped foliage that greatly reduces the availability of dissolved oxygen in the water below. This creates an oxygen-depleted environment unwelcoming to all but the hardiest fauna—such as carp, another aggressively competitive import, although one that is not yet a problem in metropolitan New York.

In sum, many of the most conspicuous organisms encountered by New Yorkers belong to alien, invasive species: pigeons, starlings, house

("English") sparrows, and Norway rats; plants such as viper's bugloss, Norway maples, and purple loosestrife; and the unseen world of disease-causing fungi and microbes. Many of the problem-causing aquatic animals and plants live out of sight below the water's surface. And yet they are vibrant and alive—and not to be disparaged when it comes to regretting the loss of so many native species. There is a certain wonderful aspect to the ability of some extremely hardy species to make a success of it in the concrete jungle.

But the story becomes even better than that when we turn to the persistence of some of our own homegrown stock—and the successful efforts of New York and its citizenry in preserving, restoring, and in a very real sense, redeeming our deeply threatened heritage of biological diversity.

CHAPTER SEVEN

Resilience, Restoration, and Redemption

If we have learned any lessons at all from the study of evolution, we humans who seem to prize stability above all else must recognize, however reluctantly, that change is inevitable. Earth's natural features—its geography, atmosphere, and climate, and all its living population, from microbes to humankind—remain ever in flux. Geneticists scanning change on the molecular level see a constant flicker of mutation and variation even while the vast, ever-mingling gene pool of a huge population such as humanity negates the present potential for further significant human evolution. And looking back on the long history of Earth (4.65 billion years) and the slightly shorter history of life (some 3.8 billion years), we see monumental changes. Continents move around, collide violently, and pull apart, creating major consequences for global climate and for life itself.

Once complex multicellular life had evolved—and especially after it began leaving a dense, rich fossil record (at the beginning of the Cambrian period, about 545 million years ago)—new species evolved and usually survived (often without changing much) for as many as 5, or even 10, million years. But there is one singular, ineluctable fact about life's history that is relevant to our narrative of the fates of species in the concrete jungle: nearly all the species that have ever lived on Earth are long-since extinct. In fact, extinction often seems to be a necessary precondition for further evolution.

As we saw in chapter 1, the current Sixth Extinction is entirely human-caused. Even the great movement of glaciers in the Pleistocene, a series of events that modified environments drastically and ended comparatively recently, accounted for the extinction of relatively few species. It was when we humans started moving freely and in considerable numbers around the globe—and especially after agriculture had become the dominant mode of making a living—that species started to die off in alarming numbers.

But our actions now are mimicking what nature has wrought in other ways during the five or six major mass extinctions that occurred in the geological past. We are the equivalent of, for example, the extraterrestrial impact(s) that brought the age of dinosaurs—and so many species on land and sea—to extinction 65 million years ago.

Given the background facts of the history of life, and the constant flux of modern genetic systems, "restoration" in the sense of a complete return to some elysian era of the American landscape is not in the cards: it simply is not possible. What's left today is postindustrial rather than pristine. Too many species are extinct, including those never documented in the first place; too much has been disturbed beyond repair, right down to the loss of the rich layer of loam and topsoil dating back to the post–Ice Age days. And, looking at New York City, the basis of a modern economy in real estate rather than in communal holdings puts the final kibosh on any restoration beyond a scattering of protected green oases.

That said, the metropolis is now enjoying the fruits of a conservation movement that began gaining critical momentum in the last quarter of the twentieth century. The Central Park Conservancy launched its heroic rescue of a much-abused and neglected landscape in 1980 with a highly successful and widely copied public-private model now applied to parks and preserves worldwide. The federal Gateway National Recreation Area, encompassing 26,607 acres of marshes and shoreline surrounding the New York Harbor, was created by congressional fiat in 1972.

Turning then to matters of resilience and redemption, we enter the realm of the possible. Here we find examples of exciting new "green" ventures reclaimed from industrial rejects such as the High Line; and here, too, we find promising works in progress, such as the conversion of a fifty-year accumulation of garbage into the verdant promise of a new Fresh Kills park—planned to "feature every environmentally correct practice known to landscape architecture." Aerial shots of New York City once revealed the occasional penthouse roof garden; now such shots show neighborhoods heavily networked with rooftop greenery, planted often as gardens, sometimes as farms, at other times for insulation and runoff control in large institutional complexes. Small, long-neglected parklands throughout the metropolis have been lovingly rescued and restored, while once-sterile islands of concrete give way to the petitions and green thumbs of neighborhood gardeners.

In a classic case history, the genesis of the two largest parks in the New York City borough of Queens is a tale of the path from raw nature to the paved roads of the industrial era, to the ash heaps of commercial production and dense human population, past a transitional phase, and on to the present mix of natural and cultural features. In addition to housing extensive educational and recreational facilities, a major-league baseball team, and a world-class tennis center, the grounds of Flushing Meadows and Alley Pond Park are tamed, reclaimed, and, if not entirely restored, at least granted the care and protection that will allow hardy species to flourish.

Overlooking Alley Pond Park, we find a series of three main parcels that starts at Little Neck Bay in Douglaston, extends south to Queens Village, and is split up by the Long Island Expressway, the Cross Island Parkway, and the Grand Central Parkway. Casting our mind's eye back to the birth of this landscape, we can see the park's underlying structure, the heaped moraine of sand and rock once embedding great melting chunks of ice left by the receding glacier of the last ice age, encountered in chapter 2. The freshwater ponds dotting the park—a 150-acre glacial trough letting off the bay—and the great boulders flanking the

Figure 67. Restoration of a 16-acre marsh in Alley Pond Park in the eastern part of Queens. The 635-acre park (Central Park contains 840 acres) boasts fresh- and saltwater marshes, tidal flats, meadow, and forest. Many of its features are associated with the glacial terminal moraine in the park. The park's environmental center contains a library and exhibits about local animals, and it sponsors education and lecture programs. Photograph by Sidney Horenstein.

hillsides to the south remain from the Pleistocene era. Today this wonderfully diverse ecosystem boasts both fresh- and saltwater wetlands, bayside tidal flats yielding to meadows, and forestation farther inland. Yet what exists now has evolved in the face of successive waves of human intervention. Following the presumably sustainable resource usage of the native Mattinecock along Little Neck Bay, colonists of the late eighteenth and the nineteenth centuries built mills along the site of the future Alley Pond Park and pursued light industry; yet the area retained much of its rural character until the early twentieth century.

By 1908, motorcars were already a fact of life. That year, William K. Vanderbilt Jr. opened the first concrete highway in America, the Long

Island Motor Parkway. Funded by subscription as a private enterprise, the road initially started in Bethpage and wound westward to cross what is now Alley Pond Park, giving Vanderbilt a limited-access road for his prized race cars to compete on. The wildly popular Vanderbilt Cup races were run on the parkway until the 1914 rally, when, after two spectator fatalities, the road was deemed too risky for racing. But tolls were pricey and maintenance costly, and the narrow roadbed with its tight curves and low overpasses was soon outmoded by more modern road-building methods.

As the parkway's condition declined over two decades, the forceful plans and political moxie of Long Island State Parkway Commission chairman Robert Moses prevailed; rather than improving the obsolete Vanderbilt road, the state built the Northern State Parkway. Construction began in 1931, and when the first section opened in 1933 the connecting Long Island Motor Parkway was finished. Ownership of the Vanderbilt road was transferred to the State of New York in lieu of back taxes, and after title was transferred to the Department of Parks in 1938, a two-and-a-half-mile bicycle path from Cunningham Park to Alley Pond Park was opened on the old roadbed—the only significant section of the old forty-five-mile parkway that remains.

Meanwhile, two opposing land-use forces were at play: in his wisdom, Robert Moses envisioned suburban bedroom communities served by car-friendly highways and access to plenty of recreational land. So even as the city bought parkland parcels such as 330 acres around the Alley in 1929, along with an option for 500 more acres for greenways to connect separate parks, much of the Queens marshland was being filled in for the sake of roadbeds and dry playing fields. It was not until 1974 that the Department of Parks created the Wetlands Reclamation Project in recognition of the integral role natural wetlands play in a healthy ecosystem. Today, Alley Pond Park's varied habitats are home to abundant bird life and a native cattail marsh that stretches across acres filled with duckweed, water plantain, and arrow arum that shelter such diverse wildlife as killifish, toads, turtles, and muskrats. A new boardwalk and nature center, surrounded by a large marsh restoration project, welcomes visitors. Restoration involves removal of invasives and reintroduction and protection of vulnerable wildflowers such as swamp milkweed and marsh St. John's wort. Elsewhere in Alley Pond Park, however, *Phragmites* (common reed) still dominates. An overview of the park's history of decline, destruction, and recovery is on display at the Alley Pond Environmental Center, opened in 1976.

Flushing Meadows–Corona Park, too, began life as a salt marsh. Facing Flushing Bay, easily accessed by water, and only a short haul from Manhattan Island, the marsh was overflowing with ashes and trash; the dump site already in use at the salt marsh was further filled when the Brooklyn Ash Removal Company contracted, in 1909, with the site's owner to dump ash there from thousands of homes. It grew into a vast, noxious landfill by the 1920s, when F. Scott Fitzgerald wrote his hellish description of the place in *The Great Gatsby:* "About half way between West Egg and New York the motor road hastily joins the railroad and runs beside it for a quarter of a mile, so as to shrink away from a certain desolate area of land. This is a valley of ashes—a fantastic farm where ashes grow like wheat into ridges and hills and grotesque gardens; where ashes take the forms of houses and chimneys and rising smoke, and finally, with a transcendent effort, of ash-gray men who move dimly and already crumbling through the powdery air."

Once nicknamed "Mount Corona," the hundred-foot-high dump had been operated since the 1920s by one Fishhooks McCarthy. In 1934, Robert Moses, who was then parks commissioner, acquired the land and managed to redirect over 50 million tons of waste for use as fill for the highways that now intersect Queens and its parks. Famously, the reclaimed land where the dump had stood became home to two world fairs—in 1939 and 1964—not to mention a swimming pool, ice rink, two museums, and a botanical garden and zoo, as well as home to U.S. Open tennis matches and Citi Field, headquarters of the New York Mets baseball team.

Nothing remains of the old intertidal salt-tolerant cordgrass *Spartina altiflora* that covered much of preindustrial Flushing Meadow. The former grasslands are mostly paved, a good deal of the Flushing River lies channeled belowground, water circulation is too sluggish to support much diversity in aquatic life, and neighborhood explorers who have scratched the park's surface report finding a layer of ash and clinkers just below—surely still polluting the groundwater. In short, this highly modified popular urban resource is far more recreational than natural. Nonetheless, Flushing-Corona's Willow Lake—twin to Meadow Lake, created as a picturesque backdrop for the 1939 World's Fair—was always meant to be a natural refuge. Recent restoration work has focused on replacing *Phragmites,* porcelainberries, mugwort, and other invader species with hundreds of new trees, shrubs, and wildflowers, all meant to attract birds, butterflies, bugs, and bigger wildlife. And in another part of the park, the Queens Botanical Garden invites visitors to Flushing Meadows to "visit the Garden's 39 acres to see native plants in bloom. Discover natives in the woodland, perennial and meadow gardens on what used to be a site operated by the Brooklyn Ash Company."

The genesis of Manhattan's High Line Park reflects a flip of the culture-nature coin. Here, evolution of the landscape shows nature overwhelming the remains of industry, rather than vice versa. The High Line opened in 1934, an elevated West Side spur of the New York Central Railroad running down from Thirty-Fourth Street to St. John's Park Terminal at Spring Street. It was a lifesaver at the time, eliminating 105 grade-level crossings and relieving Tenth Avenue of its reputation as "Death Avenue." For three decades, direct rail connections into West Side industries, warehouses, and meatpackers served the city well, but federal support of an interstate highway system, followed by the growth of trucking, shut down the lower end of the High Line by the 1960s, and after one last shipment of frozen turkeys, the remaining section of the High Line closed in 1980. From then until 2005, when CSX Transportation deeded the line to New York City, random acts of nature turned the abandoned trackway into a self-seeded greenway. It is now a much-acclaimed public park flourishing above ground level for 1.45 miles, from Gansevoort Street to West Thirtieth Street. The

Figure 68. An unrestored part of the High Line. This section will be restored in the future, but for the moment we can see what kinds of plants grow spontaneously in the city in this particular habitat. Photograph by Sidney Horenstein.

neighborhood lying below the former tracks is flourishing, too, enjoying greater, profitable foot traffic; higher real estate values; and lots of well-earned free publicity.

The greenway is an unparalleled destination for strolling, people-watching, spectacular views of the cityscape, and Hudson River vistas. It is also a marvel of ecological planning: planted with 210 varieties of plants, 161 of which are native species, it is also engineered for optimal water conservation and drainage and designed to need less irrigation and care as the landscape matures. Like any good garden, it pleases the eye at every time of year—with decorative grasses and early-blooming cultivars such as witch hazel and viburnum in winter; redbud, shad-bush, and bulbs of all kinds in spring; wild petunias, rugosa roses, sunflowers, and swamp milkweed in summer; and asters, indigo, turtle-head, and "purple bush" Joe Pye weed in fall. Although many plants

Figure 69. High Line Park. This public open space and greenway was installed on a defunct elevated rail line on the west side of Manhattan. Originally an on-the-street railroad begun in 1847, it was elevated in 1934. The structure was built down the middles of blocks so that construction would not take place on an avenue, and so that the rail line would provide direct access to factories and warehouses. Local residents were impressed when plants rooted along the line after it was abandoned. In 1999, residents began the process of saving it, taking their cue from the Promenade Plantée in Paris and creating this unusual garden paradise. The first section opened as a city park on June 8, 2009. Photograph by Sidney Horenstein.

that proved themselves as volunteers during the self-seeding era remain, far more were extirpated, and this garden of green and floral abundance is as carefully planned as the formal beds at Versailles.

The story of how the High Line went from wasteland to parkland reminds us once again of how much a handful of citizens may accomplish—given enough determination and a responsive political climate. Neighborhood residents Joshua David and Robert Hammond founded the Friends of the High Line in 1999 after they happened to meet at a community board meeting packed with West Side property owners pushing to have the old eyesore torn down. Now fifteen hundred dues-paying members strong, what started as a few feisty advocates drew in grants, public support, and cooperation from Mayor Michael Bloomberg's administration to create the park.

In 2001, in a joint project with the Friends of the High Line, the Design Trust for Public Space granted a fellowship to architect Casey Jones to research the idea of reclaiming the High Line. Jones's work indicated that restoration of the High Line would bring a net gain of tax revenue, and the city council passed a resolution advocating reuse rather than demolition. In 2002, the city invoked the 1983 congressional rails-to-trails mandate and filed a rail-banking plan with the federal Surface Transportation Board in order to preserve and reuse the High Line. The following year, an open ideas competition, Designing the High Line, garnered 720 domestic and international proposals to repurpose the High Line. For the splendid results, we may thank the landscape architects James Corner Field Operations and the architecture firm Diller Scofidio + Renfro, experts in horticulture, engineering, security, maintenance, and public art.

The park's prime movers, Joshua David and Robert Hammond, remain secretary and president of the board of directors of High Line Park and are hard at work trying to secure the last few blocks of the rail line north of Thirtieth Street—which CSX Transportation still controls—for park use rather than commercial development. In a recent interview published on the website Crain's New York Business, Mr. Hammond recalled a wonderful conversation with Robert Caro, author of a notable biography of Robert Moses, *The Power Broker.* "He said Moses would have hated this project because it's bottom-up," Mr. Hammond stated. "It started with a small group of people rather than a master plan."

Yet, master plan or no, we must work with what we have. Returning briefly to Inwood Hill Park, consider what a policy statement regarding the restitution of that park suggests: "The maintenance and sustainability of the ecosystem still depend on the basic infrastructure that is capable of survival in the given environment, which is the forest flora community itself and is the reason why the bottom up approach is preferable." This refers to 197 acres of mostly intact old-growth forest and salt marsh, where century-old native red oaks and giant tulip trees overlook massed day lilies, clumps of jewelweed, and even endangered jack-in-the-pulpit. "Bottom up" literally means starting at the forest floor to assure that only welcome species grow there. Park volunteers and employees, backed up with city, state, and federal budget support, work to replace invasive species with native plants and to fight soil erosion. Invaders such as the Norway maple grow faster than native trees and leaf out earlier, producing a killer canopy too dense to permit an understory to grow, leading to bare soil. After that, erosion is inevitable, given the degree of runoff in a mostly concrete city; and further vulnerability comes when pollutants wash in from city streets and the storm drains overflow. As with democracy, the price of a healthy landscape is constant vigilance.

While the popular Inwood Hill Park sports and picnic facilities at the northernmost tip of Manhattan are nearly loved to death, the forest and salt marsh offer refuge to 150 bird species, including hairy woodpeckers and black-capped chickadees in the forest; and in the marsh, belted kingfishers, great blue herons, ducks of all kinds—mallards, canvasbacks, buffleheads, redheads, and black—and both great and snowy egrets. But since mammals don't fly (bats excepted), this isolated patch of green holds little mammalian diversity. There are raccoons and skunks, and a population of city rodents, including a genetically interesting species called the white-footed deer mouse (*Peromyscus leucopus*).

Like their larger cousins the rats, the white-footed deer mice adjust well to the crowding, noise, and special dangers of urban settings. But white-footed deer mice prefer to stay in wooded settings and rarely wander from one greensward to another along the city streets. As a result, a number of races of the mice have evolved in various parks around the city: a study showed more differences in the DNA of these "park-isolated" mice than among all the white-footed deer mice from Queens, the Bronx, and Manhattan since their advent here after the last glaciation. This is one small but telling example of the impact of human environmental destruction on enclaves of life—as well as an example of how important enclaves are to supporting life, how resilient life is, and how quick it is to evolve given the opportunity, especially in isolated populations.

Further examples of the human effects on species' survival are not hard to find. The bottom-feeding Atlantic tomcod (*Microgadus tomcod*) develops fatal abnormalities when exposed to PCBs, but the Hudson River population is thriving in spite of the river's heavy load of PCB-laced bottom sludge from General Electric's release of about 1.3 million pounds of PCBs into the Hudson River between 1947 and 1976. In the course of perhaps only twenty generations, the Hudson River tomcod have evolved a tolerance of PCBs, to the extent that their livers can contain the highest known PCB load in a marine species. Of course, if the river recovers, the genetic advantage will surely revert to non-PCB-tolerant tomcod and wipe out the resistant population. Likewise, studies have turned up a new sort of life-form, dubbed "white stuff," from the bottom of Gowanus Canal in Brooklyn. The components of this biofilm, which is composed of bacteria, protozoa, chemicals, and other debris, appears to work in synergy to scout food, exchange genes, and produce secretions that work like antibiotics to ward off toxins in the waters. The hope is that this will lead to new drugs to combat human disease. The point here is that restoration does not lie in our hands alone; our power to mediate the forces of evolution are hardly greater than our ability to foresee unintended consequences.

One further story on the subject, again involving the ubiquitous Robert Moses, will round out this brief survey of restoration ecology in New York City. As explained by Bruce Kershner in his book *Secret Places of Staten Island:* "The Protectors of Pine Oak Woods is an environmental conservation advocacy group dedicated to the preservation of Staten Island natural resources and heritage. The organization was founded to preserve Pine Oak Woods, now known as Clay Pit Pond State Park Preserve on Staten Island's south shore. In the decades since it was established in 1976 the group has successfully lobbied for the addition of thousands of acres to Staten Island's park system[,] including the Greenbelt."

The Staten Island Greenbelt is a system of contiguous public parklands and natural areas in the central hills of Staten Island. At 412 feet, Todt Hill is the highest elevation along the Atlantic Coast south of Maine. During the latter half of the eighteenth century and well into the nineteenth century, proposals to preserve the species-rich wooded terrain were unsuccessful.

The underlying bedrock, 430-million-year-old greenish serpentine, is a chunk of the ocean floor ripped off during one of the collisions that affected ancestral North America. It was eventually covered by the Wisconsin glacier, which left boulder deposits scattered over the surface.

During the beginning of the 1960s, Robert Moses, as chairman of the Triborough Bridge and Tunnel Authority, implemented plans for a parkway that would connect Brooklyn with New Jersey via the newly built Verrazano-Narrows Bridge. His proposal, the Richmond Parkway, would bisect various park elements in the Greenbelt. Beginning in 1964, several groups of conservation activists and individuals mobilized in opposition to the plans and in fact developed plans for an alternate route. In spite of the opposition, Moses's plan went into effect and construction began in 1965. The debris from the roadway excavations was piled up on Staten Island to form what has become known as Mount Moses, the top of which affords a 360-degree view of the island. After court battles, public campaigns, and struggle after struggle, Greenbelt Park was created here in 1984. The group Protectors of Pine Oak Woods was integral to the final establishment of the park.

RESTORATION ECOLOGY IN THE MARINE ENVIRONMENT

In six locations around the city, there are newly established oyster reefs, thanks to a dedicated aquaculture program that places "spat-on-shell"—juvenile oysters grown on empty shells—in dozens of locations around the city. Here's another great idea that started out small before it grew into a multiorganization, public-private oyster-rescue mission. It began when one curious kayaker, noting the occasional presence of an oyster in the bay waters of Brooklyn, wondered how to help restore them more fully. What kayaker Bart Chezar had in mind was reestablishing reefs where they had disappeared a century or more ago. He started by immersing small cages of spats-on-shell, then moved on to a wire mesh armature holding the shells, which worked even better. "Each of those pieces of mesh would weigh eight or nine pounds when we lowered them and over 60 pounds by the time we pulled them up," Chezar recalls in a *New York Times* interview.

A proven success, Chezar's pilot effort has been adopted by a consortium of the Hudson River Foundation, the NY/NJ Baykeeper group, and the Army Corp of Engineers. Today, artificial reefs measuring about fifteen by thirty feet lie anchored at six locations: Bay Ridge Flats, Governors Island, Jamaica Bay, Soundview Park, Staten Island, and Hastings-on-Hudson, north of New York City. Each site was stocked with fifty thousand juvenile oysters reared under the care of an aquaculture class from the Urban Assembly New York Harbor School. It was a team effort all the way, drawing on the cooperation and expertise of such organizations as the Bronx River Alliance, the Governors Island Preservation and Education Corporation, the NOAA Restoration Center, and the New York City Department of Parks and Recreation. This popular program now engages many more groups of students and other volunteers, and the city presently boasts over thirty mesh-grown oyster gardens around the estuary.

The oyster reef at Dubos Point on Jamaica Bay lies near a former saltwater marsh, which was paved over with dredged fill in 1912 for a real estate project that ultimately failed. But the abandoned forty-four-acre point—like much of Jamaica Bay's thirty-nine square miles—was reclaimed by nature. It became a wildlife sanctuary when the Parks Department acquired it in 1988, and Audubon New York's drive to buffer the bay has added acreage through the years. Once again, a saltwater marsh graces Dubos Point, making this an excellent birding site, one where endangered merlins and peregrine falcons are sometimes seen. The entire compass of Jamaica Bay is a rich wetland estuary bounded by parts of Brooklyn, Queens, and Nassau County and protected from the Atlantic by the Rockaway Peninsula. Although less than half the shoreline is protected, the bay is full of wild marshland and islands. Situated along the Atlantic flyway, it's a magnet for migrating waterfowl, and over 325 species of birds, 50 species of butterflies, and 100 species of finfish are seen there. In fact, the Jamaica Bay Wildlife Refuge, located on an island in the bay, is a federally designated Significant Coastal Fish and Wildlife Habitat.

The need for remediation is ongoing. JFK International Airport abuts the bay, and in addition to the inevitable runoff from acres of tarmac and concrete slicked with fuel, oil, and a million gallons of harmful deicing fluids every winter, all twenty-six of the airport's large outflow pipes dump into Jamaica Bay. Coupled with poor water oxygenation as a result of limited water flow from the creeks and inlets feeding the bay, high levels of nitrogen breed large algal blooms on the bay's surface in hot weather—these smothering mats of rank green biomass become peppered with oxygen-deprived dead fish and crabs. Loss of the salt-marsh flora and fauna that play a critical role in marine health continues apace as more acres of marsh around the bay disappear every year.

But in 2011, under the aegis of the Natural Resources Defense Council, smaller activist groups, including the American Littoral Society and the Jamaica Bay Ecowatchers, cut a deal with New York City and the state to reduce nitrogen discharge by 50 percent by 2020. The New York City Audubon Society reports that the plan will dedicate $100 million to upgrade pollution control at four sewage treatment plants that presently discharge cleansed wastewater into the bay on good days and sewage overflow on bad ones. The city also plans to open a new $400-million sewage treatment plant by Paerdegat Basin, a channel letting into the bay. This new plant in Canarsie, Brooklyn, will retain up to 1.2 billion gallons of combined sewer overflows during heavy weather, filtering out solids such as plastic bottles and fecal matter that now discharge into the bay. Experts predict that perhaps as much as 70 percent of the present nitrogen burden will be reduced by these means.

Oysters are one natural form of nitrogen control, but nature also offers another superb natural means in the form of eelgrass, a leafy marine plant that once formed underwater meadows of waving green throughout the shallow bays and estuaries of North America on both coasts. Like the finfish and shellfish it protects and nurtures, eelgrass is sensitive to disruption and pollution, so much of the nation's eelgrass has disappeared. Jamaica Bay is just one of dozens of places on both coasts where volunteers, nongovernmental organizations, and governmental agencies cooperate to replace this critical element of littoral health. It's a continuing experiment here, where shifting sediments wiped out the first plantings, in 2009. But improved water quality promises better results for another eight thousand plants woven into burlap mats and placed off Breezy Point in 2011.

Environmentalists have also turned to another natural nitrogen-reduction tool, the ribbed mussel, which is common in New York's estuarine waters. A pilot mussel project by the New York State Department of Environmental Conservation, designed to remove nutrients and other pollutants from the waters of Fresh Creek, a tributary that flows into Jamaica Bay, placed a wire mesh enclosure full of ribbed mussels at the center of the creek—where the bivalves would otherwise never find a natural foothold to grasp—in the hope that a successful experiment will warrant placing mussels near combined sewer overflows to filter out bacteria and nitrogen before they reach the bay.

If Jamaica Bay is a work of restoration, the yet-unborn Freshkills Park in the borough of Staten Island must be a work of redemption. It will be an entirely new landscape sculpted over a landfill composed of fifty-three years' worth of New York City garbage, which had been piled upon the meadows and marshes surrounding the wide, once-navigable tidal creek Fresh Kills. The history of the creek (*kill* is the old Dutch word for "creek") includes an August 1777 raid on the British Army by rebels from New Jersey, who boated to Richmond via Fresh Kill. In the nineteenth century, the kill's tidal waters supplied power to a number of mills along its banks. Boaters liked the kill as a safe "hurricane hole," where they could take refuge when weather was rough, and there was one last boatyard renting pleasure crafts on the creek in 1955.

The dump opened in 1947 with a view to "reclaiming" marshland for development after a few years. It finally closed for good after all recoverable traces of human remains were sifted from debris trucked in from the ruins of the World Trade Center, an act of redemption in itself. What remains at the 2,220-acre park site is about half solid-waste landfill and about half lowland areas, wetlands, and open waterways. The Parks Department's thirty-year plan envisions another "people's park" created with the same type of grand vision that yielded Manhattan's Central Park over 150 years ago. But this one will be three times as large and up to the latest green technology standards, "an unusual combination of natural and engineered beauty," the Parks Department predicts.

The very real differences between creating Central Park and establishing Freshkills are both political and terrestrial. True, both plans were adopted after design competitions, but in today's far more open and communicative New York, the running drew vastly more than thirty-three designs, and the reigning park commissioner did not win the competition. Fredrick Law Olmsted, senior partner of the team of Olmstead and Vaux, who designed Central Park, had been appointed park superintendent in 1857. The winning greensward bid was not judged until after the final submission, on April 1, 1858. The Central Park design called for a dramatic reworking of what remained an essentially natural landscape—notwithstanding the African-American community of Seneca Village, which included 270 or so members and three churches, plus some 1,300 other residents who were evicted from within park boundaries. Further, while consideration of the wishes and interests of powerful New York City citizens was allowed to moderate the plan somewhat, the creation of Central Park remained essentially a top-down operation.

In contrast, Freshkills represents the integration of some relatively unspoiled acreage with an entirely unnatural landscape. Moreover, the needs and opinions of citizens of Staten Island and all New York City are being much more effectively integrated into the plan than those of citizens in earlier centuries were incorporated into previous plans. When Central Park opened, it mainly served the wealthy residents of uptown mansions and the middle-class visitors who could afford the carriage or train ride up from town. Residents of crowded tenements, who most needed the relief of a green retreat, found Central Park too far to reach on foot and too costly to ride to. On Staten Island, a fully realized Freshkills Park will be more than just the strongest link in a greenbelt winding through the island. It will also meet the demands of sports lovers and nature lovers, who want to play soccer, ice-skate, ride trail bikes, and kayak, and who want the park to include a variety of amenities, from hardened playing fields to a native-plant garden and environmental center.

What we see here—and around the world—about five generations after New York's city fathers created the first grand American public park, at the heart of Manhattan, is nothing less than a total paradigm shift. A look back to that past just before the Civil War reveals a hierarchical, linear society where decisions rained down from above. The spread of information was limited not only by many people's inability to read a line of print and by the absence of transparency among the powerful, but also by poor access to a slow and restricted flow of communication, compared to today's utterly networked global society. In fact, our modern concept of ecology views nature, too, as a web—a network of life with which we humans try to mesh our needs, with a maximum of remediation and a minimum of disruption.

During the planning stages of Freshkills Park, a summary of community expectations included wishes to keep the site passive and natural; maintain access to water; have large-scale open spaces with paths and trails for long walks, cycling, and horseback riding; allow site-limited commercial activities only at the core of the park; provide plenty of sports and recreation facilities; and offer educational interfaces to highlight renewable energy and ecological techniques of land reclamation. All these goals are now folded into the master plan, which is expected to take until at least 2040 to fulfill. The first stage is reaching completion now at Owl Hollow: it includes the installation of four synthetic-turf soccer fields and a comfort station certified by the U.S. Green Building Council through its LEED (Leadership in Energy and Environmental Design) program. The comfort station is complete with a green roof, geothermal heating and cooling, and a wind turbine that will power the building. In March 2011, city officials were joined by members of the Staten Island community to commemorate the tenth anniversary of the landfill's closing. They greeted a barge loaded with trees and conducted a planting ceremony in South Park, the first of the five distinct areas that will constitute Freshkills Park: the Confluence, North Park, South Park, East Park, and West Park.

In light of the public process behind Freshkills Park, it is interesting to note that Central Park abounded in keep-off-the-grass signs up until World War II, and the first children's playground was not opened there until 1927. By now, we're approaching the half-century mark of heightened environmental awareness and activism in the metropolitan area. The concept of biophilia, meaning a love of nature and animals, was first floated by Erich Fromm in 1964 and later made famous by Harvard biologist E. O. Wilson. At first, it was an idea honored more in the breach than in the observance, since the city's public spaces were displaying the effects of a neglect bred of social ills, overuse, and underfunding. But that time also saw the birth of a "power to the people" movement when a fresh sense of individual empowerment and responsibility swept the nation.

While some were protesting the war in Vietnam, others were planting gardens. In 1973, Liz Christy, an artist living on the Lower East Side, gathered a group of friends to clean out a rubbish-laden city plot at the corner of Bowery and Houston Streets. Becoming activists, they dubbed themselves the Green Guerillas, who threw "seed bombs" into vacant lots, planted window boxes and tree pits with flowers, and prevailed upon the city to designate the first New York City community garden, named after Liz Christy herself. That groundbreaking effort, originally called the Bowery Houston Community Farm and Garden,

Figure 70. The first community garden planted in New York City, founded in 1973 on Houston Street, between Bowery and Second Avenue. The garden was named after Liz Christy, the driving force behind its creation, but it has been known by other names, too, including simply the Community Garden. The site, a treeless vacant lot, was converted to a secluded place in the teeming city with lovely paths and a great variety of flowers. It's also home to the largest dawn redwood (*Metasequoia glyptostroboides*) in New York City. Planted sometime during the late 1970s, the tree now soars to a height of more than one hundred feet. Photograph by Sidney Horenstein.

has grown into a thriving 501(c)(3) nonprofit organization, and today the Green Guerillas—now more formally known as GrowNYC—point with pride to the more than sixty community gardens around the city in which they've had a hand.

Similarly, the New York Restoration Project is committed to land stewardship, owning and managing fifty-five community gardens formerly destined for sale to developers. It all started with the return of a grateful diva to the city where she got her start. Bette Midler was horrified to find Fort Washington and Fort Tryon Parks reduced to filthy wrecks in 1995, so she started cleaning them up with the help of friends and family, an effort that grew into the present nongovernmental organization. Now the New York Restoration Project works closely with the New York City Department of Parks and Recreation and an army of corporate supporters and volunteers. The latest collaboration is the Million Trees Project, one of 132 PlaNYC 2030 initiatives launched under the auspices of Mayor Bloomberg in 2007. At this point, the project is more than halfway completed, although the care and nurturing will continue.

CANYONLANDS AND THE FUTURE

Many people who have visited the city in the last decade or more marvel at the pace of building that has taken place. Two- and three-story "taxpayers" and five- and six-story brownstones have been replaced by massive buildings. As I walk around the streets and look at these new structures—high-rise condominiums, office towers, superblocks, and even structures enclosing football fields, which have incrementally swallowed more and more of the sky—I muse about what the future will hold, and about how much more of the sky will be blocked. While natural erosion tends to reduce geological canyonlands, the canyons of Manhattan seem to proliferate and grow ever deeper as the buildings grow taller. For people with a touch of claustrophobia, the prospects may not look good. But for the greater majority of people, this is what the city of the future should look like.

We may find a hint of the future in PlaNYC 2030, issued by Mayor Michael Bloomberg in May 2007. In summary, this plan assumes that the city's population will increase by 1 million residents by 2030, and it presents a way to accommodate all these people. The plan is comprehensive, rather than haphazard, and it contains initiatives to strengthen the economy, control climate change, and enhance the quality of life for all New Yorkers. To fulfill these goals, the plan envisions ten areas of interest that must be addressed: housing and neighborhoods, parks and public spaces, brownfields, waterways, water supply, transportation, energy, air quality, solid waste, and climate change.

After the plan was released, 127 initiatives were proposed, and by 2009 many of these initiatives had been implemented. The plan was updated in 2011, increasing the number of initiatives to 132 and listing more than 400 milestones (the steps taken to accomplish those initiatives) to be completed by December 31, 2013, the end of the mayoral term. Readers can access the plan at www.nyc.gov/html/planyc2030/html/home/home.shtml, which contains the full updated report issued in April 2011. Many groups and individuals have responded to the plan, generally offering accolades but also some criticism. Some of these responses can be found in newspapers, such as the *New York Times*, and elsewhere on the Internet by searching for "PlaNYC 2030."

Let's look at a few of the goals proposed in the plan. One of its components is a reduction in our reliance on landfills by converting solid waste into energy and heat. Although to some people, this means burning garbage, as well as the release of unwanted particulates and gas, the plan mentions that the city may explore technology that converts waste into energy. For now, the plan entails the use of out-of-town landfills. All of the city's landfills have already been closed; and long before that happened, ocean dumping ceased. Sending garbage to distant places is not a long-term solution— eventually these dumps will fill up and the number of cheap sites will be diminished, depending on local laws.

Although usable technology for converting garbage to heat and electricity exists elsewhere—for example, in Denmark—no plants equipped to do so currently exist, or are being planned or built, in New York City or anywhere else in the United States. Landfills are cheaper right now, even though the city must pay to truck the 10,500 tons of residential garbage generated each day to landfill sites in states such as South Carolina and Ohio. People who object to building garbage-conversion plants say that these would be the demise of community recycling programs. They propose 100 percent recycling so that such plants would no longer be needed. It should be noted, however, that some countries with garbage-conversion plants, such as Germany, have the highest recycling rates.

With the city's projected increase in population, housing is an immediate concern. Here, the goal is to preserve as much affordable housing as possible and create new affordable homes. This can be accomplished by rezoning initiatives and recycling older commercial buildings for housing that are no longer appropriate for their initial use. As available space for new buildings diminishes, the city could put decks over rail yards and, in some cases, over highways and build on top of these decks.

One of the apparent visual improvements in the city is the creation of green streets, where dividers and median malls have been planted with flowers and shrubs. Although the Greenstreets program has nearly reached its goal, the larger, related initiative calls for planting 1 million trees in the city. Eight hundred thousand trees had been planted by November 14, 2013. In addition to new parkland—the city has recently added 182 acres' worth—the plan calls for opening schoolyards as local playgrounds, with the goal that no New Yorker will be more than a quarter-mile walk away from a park.

The goal of planting trees is to make the city visually more pleasing, but also to help clean the air by reducing carbon emissions. Like all coastal cities, New York is vulnerable to the threat of sea-level rise as a result of climate change. Many of the initiatives are interrelated in their efforts to slow down climate change. They propose reductions in emissions from power plants and from automobiles (by improving mass transportation and thereby reducing the number of cars that enter the city), as well as adding "green" buildings and increasing the number of hybrid taxis and hybrid city vehicles. But as helpful as the proposed programs are, the city is still dependent on emission reductions around the globe and upon how much sea level rise is natural. As a result, PlaNYC 2030 calls for a long-term effort to develop a comprehensive plan for climate shifts that are probably unavoidable. Everyone is aware that the city will change, but PlaNYC 2030 gives focus to the change and indicates the direction New York City will take in the future.   *SH*

Figure 71. The Saturday Greenmarket on Isham Street, in northern Manhattan. The Greenmarket program was created in 1976 to promote regional agriculture by providing small family farms with the opportunity to sell their locally grown products directly to consumers. Today, there are about sixty-seven such markets, featuring over 230 family farms and fishermen. These farms represent thirty thousand acres of farmland protected from development. Photograph by Sidney Horenstein.

Returning to the subject of bottom-up activism, an overview of the entire metropolis today reveals a network of green nodes—an ever-increasing collection of gardens planted by New Yorkers. Throughout the city, bare roofs and abandoned lots have grown into farms and gardens in synch with the advent of greenmarkets. The first farmers market opened at Fifty-Ninth Street and Second Avenue in 1976. Today there are more than sixty seasonal and year-round greenmarkets in New York City, all of them offering local, often organically grown, fruits and vegetables, as well as honey gathered from resident bees.

Besides promoting healthy eating, community gardens offer hands-on experience to the public, especially to schoolchildren. More than eighty rainwater reclamation projects now under way at community gardens are cutting down on storm runoff and promoting conservation awareness. The Brooklyn Grange farm, founded by New York Botanical Garden expert Annie Novak atop a former bagel factory in Long Island City, Queens, is a spin-off of the first New York City nonhydroponic rooftop soil farm, Eagle Street Rooftop Farm of Brooklyn, founded in 2009 by Wisconsin native Ben Flanner and Novak. Eagle Street Rooftop Farm was the city's first community-supported agricultural program. Flanner and Novak now claim to operate the world's largest rooftop garden and Brooklyn Grange recently expanded to a second near-acre of rooftop in the Brooklyn Navy Yard.

Corporations in the city are both backing community-supported agriculture and adding green technology to their own bottom lines with green roofs that conserve energy by means of solar panels and insulation and by recycling water. New York City boasts the first-ever green apartment tower: the Solaire in Battery Park City, a gold-level, LEED-certified site, includes rooftop photovoltaic panels said to reduce peak demand by 65 percent. The building also uses 50 percent less potable water than conventional high-rises.

Thus, a network of greenbelts and blue waters emerges as biophilia becomes public policy. In every borough of the city, agencies and organizations are working to restore access to cleaner watersides and to open connecting paths to skirt city congestion. The Staten Island Greenbelt connects parklands from the central forested High Rock Park to Freshkills Park. In the Bronx, the Bronx River Greenway connects to the Mosholu Greenway through the Bronx River Forest, and connects to a greenway in Van Cortlandt Park and exits the park on Broadway. In Manhattan, the Hudson River Greenway, the East River Greenway, and the Harlem River Greenway present a thirty-two-mile path for walkers and cyclists along the waterfront, although there are some long breaks in continuity and places where the walking and bicycle paths diverge.

Much remains to be accomplished: the Greenbelt in Brooklyn is a high-end apartment complex advertising eco-friendly "sustainable"

Figure 72. A remnant of the Vanderbilt Parkway, in Alley Pond Park, Queens. The parkway, also known as the Long Island Motor Parkway, was a toll road built in 1908 by William K. Vanderbilt Jr., and the first roadway designed exclusively for automobile use. Only forty-five miles of the planned seventy-mile length was constructed. The road was closed in 1938, and most of it was destroyed by suburban growth or incorporated into newer roads. A few sections still remain. Photograph by Sidney Horenstein.

living, while the Queens Greenbelt Park is the old Vanderbilt Parkway greenway. As part of its environmental outreach, the Bloomberg administration launched a "blue network" campaign to restore the waters flowing in and around New York City. Much of the emphasis—on improved sewers, reduced runoff, and solid-waste reduction—serves the goal of protecting the waters lapping the city's 520 miles of shoreline.

The Hudson River Estuary, crowning jewel in New York's blue necklace, shows signs of recovery, along with signs of yet more exposure to leaching brownfields caught between remediation and commercial waterfront development. One example on the New Jersey side is the Quanta Superfund site at the shore in Edgewater; the Environmental Protection Agency plans to remove four feet of soil, then use fresh soil or asphalt or concrete to cap the fifteen-acre site polluted by a century's worth of coal tar and PCB-bearing oil. When completed, the twenty-one-mile, thirty-foot-wide Hudson River Waterfront Walkway, mandated by the state of New Jersey, will pass through Edgewater as it runs from the Bayonne Bridge south of Hoboken north to the George Washington Bridge, where the Palisades Interstate Park trails begin. The plan requires all developers who build along the Hudson River Waterfront Walkway to incorporate this public space into their footprint—at least in theory; whistleblowers find that, on the walkway, about 50 percent of the "public" space in private hands is locked or otherwise inaccessible to visitors.

On the Manhattan side, between 137th and 145th streets, the North River Wastewater Treatment Plant opened in 1986. Since then, for the first time in New York City's history, raw sewage is no longer being discharged directly into the Hudson River. Barges transport the liquid sludge to the Wards Island Wastewater Treatment Plant, where it is dewatered.

There was plenty of initial community objection to the location of the North River Wastewater Treatment Plant. It turned out that the plant frequently emitted noxious odors into Riverside Park and the densely inhabited Harlem; $55 million dollars later, the problem seems to have been solved. The bright side (apart from solving a major pollution problem in the Hudson) is the creation of Riverbank State Park, opened in 1993 on the roof of the water treatment building. The park attracts up to ten thousand people a day—making it the third-most-utilized park in the entire New York State system. Inspired by massive rooftop gardens in Tokyo, the park is landscaped with flowering trees and other plants, in addition to featuring well-used recreational facilities.

The Hudson River itself is cleaner than in living memory; the blue crabs and shad again occupy ancestral breeding grounds in the estuary. The New York Department of Environmental Conservation announced

Figure 73. The massive North River Wastewater Treatment Plant, in the distance. Built on the Hudson River, and reaching from 135th to 145th Streets, the plant treats 350 million gallons of wastewater and runoff daily. Some of the pilings it rests on go down as deep as one hundred feet. Photograph by Sidney Horenstein.

that "construction of sewage treatment plants and control of pollution discharges have led to water quality improvements all along the Hudson. Environmental laws, education and advocacy have worked together to improve the Hudson and surrounding land. Not only have birds and fish benefited, but today people can enjoy a cool swim on a hot summer day at river beaches, and can fish, boat and watch wildlife along the river's length." All this is true, but as we write, General Electric is expected to continue, for at least five years, the final stages of PCB dredging in the upper estuary, and swimming advisories may be in order.

In 2001 the state Department of Environmental Conservation laid out its blueprint of how to help the Hudson River Valley ecosystem, pointing out that personal commitment and a willingness to become politically active are keys to success. These recommendations serve as a summary of the current remediation efforts and of the potential for other efforts in New York City. The suggestions include the following:

- Protect large, contiguous, unaltered tracts wherever possible;
- Preserve links between natural habitats on adjacent properties;
- Preserve natural disturbance processes such as fires, floods, tidal flushing, seasonal drawdowns, and wind exposure, wherever possible;
- Restore and maintain broad buffer zones of natural vegetation along streams, along the shores of other water bodies and wetlands, and at the perimeter of other sensitive habitats;
- In general, encourage development of altered land instead of unaltered land wherever possible;
- Promote redevelopment of brownfields, other post-industrial sites, and other previously altered sites, instead of breaking new ground in unaltered areas;
- Encourage pedestrian-centered developments that enhance existing neighborhoods, instead of isolated developments requiring new roads or expanded vehicle use;
- Concentrate development along existing roads and discourage construction of new roads in undeveloped areas. Promote clustered development wherever appropriate, to maximize extent of unaltered land;
- Direct human uses toward the least sensitive areas, and minimize alteration of natural features, including vegetation, soils, bedrock, and waterways;
- Preserve farmland potential wherever possible;
- Minimize areas of impervious surfaces (roads, parking lots, sidewalks, driveways, roof surface) and maximize onsite runoff retention and infiltration to help protect groundwater recharge, and surface water quality and flows; and
- Restore degraded habitats wherever possible, but avoid destruction of existing intact habitats.

In closing, on the subject of existing intact habitats, let us recall that a "weed" is nothing but a plant uninvited by the gardener, and recall, too, that nature cares nothing for the fantasy of a pristine environment. When Peter Del Tredici, senior research scientist at the Arnold Arboretum of Harvard University, looks at invasive urban plants, such as *Ailanthus,* Japanese knotweed, and *Phragmites* reeds, he sees hardy survivors capable of covering brownfields, refreshing swamps, scrubbing and cooling the air, mitigating runoff, and beautifying the landscape in spite of utter neglect and harsh conditions. Interviewed by the *Boston Globe,* Del Tredici said, "I consider 'weed' to be a politically incorrect term. There is no biological definition of the term weed. It's really a value judgment."

And so, concerning restoration, resilience, and redemption, it is remarkable how the mounting environmental awareness over the last few decades of the previous century, which has steadily gained force in the new millennium, has led to an unprecedented determination to make things better. Cherishing natural habitats and species leads to active programs and commitments to conserve and restore. The world has never seen the likes of this before. The consciousness of the need to restore and conserve has never been higher—and this is perhaps truer in cities like New York than it is in many other, more rural, and therefore superficially less endangered, environments. And nowadays, the agents for turning the environmental tide are found as frequently at the grassroots level as at the politically powerful top. No longer can a Robert Moses, for better or worse, call all the shots.

But given the global economic outreach of cities, as well as the local impact of people everywhere, cities jointly have a constantly increasing deleterious effect on virtually every square inch of the surface of the globe, not just in the cities themselves. Yet cities, given the newfound power of the environmental movement, are emerging as the last, best hope of global ecosystems—and of the fragile networks of species that remain alive.

CHAPTER EIGHT

Cities, Globalization, and the Future of Biodiversity

Cities are the apotheosis of environmental destruction. They can be cleaned up, and to some extent their natural settings can be restored. Pockets of original habitat can still be found in cities, such as the acres of midwestern prairie land found between abandoned railroad tracks in Chicago, and the native species still found in Inwood Hill Park in New York. Native animals and plants regularly—and apparently increasingly—reinvade old haunts that are now converted to the concrete jungle. No city represents a 100 percent environmental transformation. But most cities come pretty close. As the world's human population expands, and the migration to the cities intensifies (for that's where the money is) cities will most likely become more and more the place where most people live. Those cities will become more and more vertical and exclude just about every other animal species, except the hardy commensals like rats and house sparrows, who earn their often rich livings from the leavings of their citified human hosts. Nearly half of the world's 7 billion people currently live in and around cities.

Conserving biodiversity is a global issue—and while the maxim "think globally, act locally" is exactly the right idea, most preservation of the world's remaining biodiversity necessarily takes place where large chunks of at least semipristine wilderness remain. So how could it possibly be true

Figure 74. Dead rat on a bluestone path in Isham Park, northern Manhattan. The number of rats in New York City is uncertain, but it is often said that more rats than humans live here. Known from early colonial times, the Norway rat, or brown rat (*Rattus norvegicus*), is generally nocturnal and lives off of, and alongside, humans. This rat may have died of old age or poison bait; the population in the city fluctuates according to the availability of food and shelter and in response to efforts to control it. Photograph by Sidney Horenstein.

that cities are the last, best hope of the world's ecosystems and species? The short answer: because that's where the money is. In this instance, that means political and financial institutions, educational and research institutions, large-scale media operations, and nongovernmental organizations (NGOs). New York's mayor Michael Bloomberg once famously remarked, when complaining aloud that his city did not receive its fair share of post-9/11 antiterrorism money from the federal government, that you don't find terrorists in cornfields. And indeed, the targets attacked on that horrendous day say it all: the Pentagon, symbol of U.S. might around the world; and the World Trade Center's twin towers, icons of globalization. The fourth plane reportedly was headed toward the White House.

It is in cities that the concentration of power—wealth, governmental agencies, media, NGO headquarters, zoos, museums, and most educational institutions—in largest measure is found. But the outreach of cities has been more often destructive than constructive. Just as New York (and almost all other cities—Los Angeles leaps to mind) depend on the watersheds of their regional surroundings (often, as in the case of L.A., to the detriment of those watersheds and of the people outside the city, who also need that water), city dwellers, by dint of their sheer numbers, have a consumer's reach that has long been global.

The Panama Canal is a century old and is currently under expansion. And although enough of Panama's rain forests were spared—so that rainfall is generally adequate to keep Lake Gatun sufficiently filled to supply the 55 million gallons of water needed to flush each ship through the locks—the enormous amount of physical destruction, and of course loss of human life, entailed in constructing the canal was a simple by-product of the desire to get coffee and thousands of other commodities more quickly, hence less expensively, to market. And most markets are in cities like New York, where those large ships tend to dock.

Preservation and conservation are a tough sell, perhaps especially so to city dwellers, who tend not to know where their food, water, and electricity come from. The very nature of the human experience, at least for the vast majority of us, centers on, even relies on, our divorce from the natural world. This divorce is responsible for the finest aspects of human cultural existence—museums, libraries, concert halls, universities, and other vital, if less unambiguously wonderful, institutions of the financial and political spheres.

As we saw in chapter 1, cities began as an almost inevitable consequence of the invention of agriculture: agriculture demands a settled existence. With the advent of agriculture, humans became the first species to take life into its own hands—to take charge of food production. Despite the famines that still bedevil huge numbers of people, our agricultural efforts have been so successful that the global human population has mushroomed, from the approximately 5 million people believed to have been on earth ten thousand years ago, when agriculture began to emerge as a serious strategy for human existence, to the more than 7 billion here today.

With agriculture, humans began to live outside of the immediate context of their local ecosystems. Native plants, especially the fast-growing, hardier sorts, became "weeds" as they tried to hang on, competing for space, nutrients, water, and sunlight with the newly minted crops that were the initial outcome of the earliest and crudest forms of genetic engineering: selective breeding by trial and error. At worst, the environment, including its wild denizens, became something that needed to be conquered.

Or exploited: fisheries are the biggest vestige of our hunting past. Most of the dozen or so global fisheries are severely depleted. Most sushi restaurants (in fact, most restaurants of any sort) are in cities. And the angst that besets many diners selecting bluefin tuna in a Japanese restaurant—an angst felt especially by sophisticated consumers who do know where their food comes from, and who know that bluefin tuna are in serious danger of utter extinction—is generally not sufficient to deter those diners from the undeniable delights of ordering high-quality tuna sushi.

Alongside the intensified exploitation of so-called natural resources, and alongside the view that the continued existence of primordial inhabitants is a downright nuisance, lies the neglect of the natural world that is a logical extension of the fact that we live, in a very real sense, outside the natural world. Nature has become irrelevant to the vast majority of people. This is, of course, especially true of city dwellers—which is to say, most people.

But there are other reasons why conservation and preservation are such a hard sell. Although rigorous psychological studies, so far as we know, have not yet addressed this issue, it seems likely that most humans are incapable of thinking into the future beyond the next one and a half, sometimes two (and rarely three), generations. Most people who have kids live to see them grow into adults. They may also see their grandchildren growing up, but few live long enough to see their grandchildren repeat the cycle. This suggests that instant gratification is a built-in criterion governing our choices, including our choices about what we will do for the future. Setting aside some money for grandchildren to go to college is common enough. But only very wealthy, and perhaps megalomaniacal, people try to control what happens to their money further down the line than their own grandchildren. The effect of the selfish gene (said to motivate organisms to care for distant kin in proportion to the shared amount of genes) reaches only so far down the family lineage into the future in the conscious minds of most human beings.

The inability, or unwillingness, to think that looking further down the road would be a good idea has obvious implications for the topic at hand. Biodiversity loss at the hands of humanity is, in the last analysis, a reflection of the still-out-of-control human population growth—itself a reflection of our settled and now mostly citified lifestyle triggered by the invention of agriculture. If most people cannot think beyond their own grandchildren (if they have any), the future state of human population growth is of no immediate concern to them.

But this myopic, even if natural and understandable, restriction of our gaze to our own lives, or to our lives and those of our already-living descendants, has an even more immediate effect. It dampens what little ardor might exist for the conservation and preservation of the natural world, including the world's still-living species, since conservation and preservation come at a cost—a direct economic cost. Or this is how conservation and preservation are perceived. Much of the resistance to accepting the plain fact of global warming comes from political and economic sectors who fear the implicit threat to the old way of doing business: of thinking of the earth as a huge (if shrinking) repository of natural resources put there for our use. And many people fear that it will simply cut too far into profits to rethink—to redesign and reengineer—our approach to power generation, our reliance on fossil fuels for transportation, and so forth.

This fear is not entirely misplaced, because it often takes a while for the benefits derived from ecologically sound solutions to become apparent. Hybrid cars are increasingly acceptable because, though their initial costs may be a bit higher than comparably sized and equipped cars with internal combustion engines, they cost less to run: they get better mileage. Thermopane windows have become worth the installation costs because they save on heating bills. And the McDonald's fast food chain is said to have switched to cardboard from styrofoam packaging because it was pointed out that recycled paper products actually reduced their packaging costs.

We still need the world's biodiversity to maintain our own quality of life. Of the 7 billion people on earth, something like 3 billion, maybe more, do not have ready, reliable access to safe drinking water. For renewable sources of safe water, we need our forests. And protecting our fisheries much better than we have been doing is essential to the continued existence of this valuable protein resource. And beyond maintaining healthy ecosystems around the world for the sake of prolonging the availability of the natural resources that we still count on—as vital to the urban dweller as to people who live closer to nature—there is also the pure aesthetics of experiencing healthy natural ecosystems. That's why we have parks in cities. And urban kids who have never seen a cow—or a deer outside a zoo—usually take to the woods, fields, lakes, and seashore as naturally as if they had grown up with them. Ask anybody lucky enough to take advantage of the *New York Times* Fresh Air Fund—which has, for well over a century, been treating some underprivileged New York City kids to a few weeks in the rural outdoors— what they think about that experience.

In a nutshell, cities are more than superficially the diametric opposite of the natural world that they have supplanted and, by dint of their rapacious outreach, that they continue to exploit and consume, to the detriment of the entire world, including its biodiversity. Yet it is the resources of cities that can redress the imbalance. They can do so by raising citizens' awareness about their collective impacts and by formulating practical policies and actions that can mitigate the negative impact of cities. And by finding the necessary political will—of those who are engaged in politics, finance, science, and education—to play a direct, positive role in stopping what has been up until now the primarily unchecked destruction of global nature. The key to all this is the cooperation of the political, financial, scientific, and educational sectors—perhaps, above all, the key governmental institutions and NGOs. The latter are increasingly important as governments and national economies flag.

BIODIVERSITY: SCIENCE AND POLITICS

Any attempt to redress the balance and stave off further apocalyptic consequences of the current mass extinction must be seen as inherently political—perhaps even more than scientific. It also seems to be the case that political conservatives, though often associated with antienvironmental attitudes, are by no means unilaterally, philosophically, or even religiously necessarily or automatically opposed to concerted efforts to save the planet. This means the physical environments of Earth's soils, freshwater, atmosphere, and oceans—and all the terrestrial ecosystems, home to the estimated 30–40 million species of microbes, fungi, plants, and animals still populating the planet. Conservatives are no more philosophically inclined to trash the planet than liberals are necessarily philosophically inclined to do anything about saving it.

Ask scientists, and they are almost sure to tell you that science is inherently apolitical: the only "left versus right" in science emerges from patterns of protein folding, "left brain versus right brain," and a few other, manifestly truly apolitical problems involving issues of symmetry. Yet probed further, most scientists agree that there are consequences that flow from scientific results—most obviously in practical applications flowing through engineering into medicine, agriculture, and virtually every other cultural facet of the modern world: technology flows primarily from science. And though this flow in general does not seem especially political, it immediately engenders political issues and problems, some of which are directly connected with biodiversity issues.

Consider the disparate distributions of cell phone and Internet access in the United States (as a microcosm of the entire world): In cities, there's no problem finding service, although wealthy neighborhoods tend to have more extensive and reliable access to broadband (and wireless) connections. In many rural sections, though, it is still dial-up for most who would explore the Internet, although satellite feeds provide a partial solution. And many of the least-wired portions of the United States are in protected areas—such as the Adirondack Mountains of northern New York State, where a fair measure of this book was written. It is a political dilemma, though, as those who fought so long to keep the Adirondacks as nearly pristine as possible—natives, to be sure, but also summer folk, many of whom, and whose ancestors, have been coming up from the New York metropolitan area over the past 150 years—writhe at the very idea of seeing "Frankenpine" cell phone towers installed in the vast stretches of boreal forest that is the heart of the Adirondack landscape.

But that is just the tip of the iceberg representing the deep connections between science and politics. Increasingly throughout the latter half of the twentieth century, U.S. science, and the academic world more generally, is funded through governmental agencies such as the National Science Foundation, National Institutes of Health, National Aeronautics and Space Administration, Office of Naval Research, and so forth. And on the face of it, this appears to be a good thing, that science is alive and well in the United States in the twenty-first century.

Yet there is a downside to this strengthened relation between science and government. Universities more than ever depend upon governmental support strictly for their survival. And government funding of course means politics, even when economic times are good, but perhaps even more so when times are tough, as they have been during the first decade of the twenty-first century and will likely be in the foreseeable future. The electorate always squabbles about where money is to be spent. The late William Proxmire, a senator from Wisconsin, was fond of periodically designating the winners of what he dubbed the

"Golden Fleece Awards." Among his favorite targets were studies on sexual behavior (in humans, but also on a wide assortment of other species) funded by the National Science Foundation and National Institutes of Health, along with many other investigations the senator thought pointless and trivial. (Most recently, and closer to home for us personally, a Tea Party candidate railed against the U.S. government for funding American paleontologists in the effort to collect fossils in foreign lands.) To be fair, given that the pot of available money is finite, some people would legitimately prefer to see their money spent on more obviously practical programs, such as food stamps, road repair, health care, and national defense, than on the sex life of quails.

But the connections between politics and the funding—and even practices—of certain forms of science go deeper than someone's opinion on what constitutes wasteful spending. Objections to stem cell research, linked as that subject is with the incendiary abortion debate, tend to come from the political right. On the other hand, the political left tends to object to genetically altered fruits, vegetables, and animals headed for the supermarket. Those luscious strawberries now available in many cities throughout the year reflect not only better transportation and preservation technology but also altered genes. And the corn that no longer turns into a chewy, starchy mess right after you pick it—that too is a product of designer genes. The political objections to genetically altered foods are inspired by moral sensibilities (and the fear that some have of the unintended spread of altered genes to wild species), as are the objections some have to stem cell research.

In the end, the concerted outreach from cities to the world's simultaneously endangered ecosystems and species cannot come initially from, and cannot be supported mainly by, governmental initiatives. The American electorate has been split pretty much down the middle on most issues of any consequence for quite some time. The 2012 removal of timber wolves from the endangered species list is emblematic of recently renewed efforts to open up vast tracts of natural land under governmental control, seen by many as necessary for American economic self-sufficiency and strength. Others see it as a simple power grab at the expense of hard-won proenvironmental battles of the not-so-distant past. Government must nearly always be involved, however, in major, concerted conservation initiatives, for the simple reason that it is governments—national, regional, and local (and even international, as in Antarctica)—that either own or exert control over the territories that need conservation programs. But conservation initiatives usually must arise from, and be sustained by, other sorts of agencies—agencies that primarily are located in cities.

There has been a great expansion in the number of successful international efforts to set aside large tracts of land, frequently with connecting wildlife corridors, and often in the name of saving a few iconic species, such as African black rhinos and Indian Bengal tigers. Each of these initiatives represents cooperation between scientific research institutions (universities, natural history museums, and zoos), political bodies, and funding agencies. At the center of such initiatives is usually one or more nongovernmental conservation organizations, several of which have grown to an impressive size over the last few decades. These organizations possess formidable economic resources and political clout. NGOs such as the International Union for Conservation of Nature (the world's first global environmental organization, founded in 1948 as a consortium of more than a thousand NGOs and governments), the World Wildlife Fund, Conservation International, and the Nature Conservancy, among others, have taken the lead in identifying conservation projects. And they have moved them forward by establishing ad hoc networks of relevant governmental agencies, scientific institutions, and funding sources from the private sector (individuals, but also foundations, banks, and other financial institutions). Conservation NGOs take the lead in such initiatives but also focus on refining policy issues and strategies, the better to maximize their spending power.

All such NGOs began in cities, and their headquarters remain in cities today, despite the global reach of the Internet, which would theoretically allow their headquarters to be located anywhere. Cities also remain the places where scientists, politicians, and moneyed interests are concentrated in critical numbers. Conservation International apparently began in a hotel room in Washington, D.C., in 1987, where a number of conservation-minded people, most of them scientists, convened and got the ball rolling. The organization is still headquartered in Washington, although its website, like the websites of most such organizations, stresses its regional and global presence and obscures the location of its headquarters. The World Wildlife Fund, too, began, and remains, in Washington, D.C.

In New York, the two major conservation initiatives with global impact are not, however, among the free-standing NGOs. One, the Wildlife Conservation Society (formerly the New York Zoological Society), runs the zoos and aquaria in the city, including the justly famed Bronx Zoo. Zoos have been much maligned over the years for keeping animals on display, often in wretched conditions, merely for the entertainment of the masses. Yet zoos play a vital role in educating city dwellers and out-of-town visitors, often providing the only look people ever get at living, breathing animals, increasingly in well-designed compounds emulating their natural habitats. It is no mean trick to make an outdoor compound in Bronx, New York, look like gorilla habitat in the Congo—but the Wildlife Conservation Society manages to pull it off pretty well. In contrast, the gorilla diorama at the American Museum of Natural History has a meticulous re-creation of the actual habitat of a mountain gorilla group, including plants collected from mountain gorilla habitat and preserved, along with accurate paintings of the background superbly rendered. Yet, of course, the gorillas are dead—so zoos and museums are not the same experience but tend to complement one another.

As a measure of how widespread global habitat destruction has become just since the 1920s, consider this: all but one of the dozens of dioramas at the American Museum depict actual locations, and recent photos taken of those locations, including the area where the gorilla group originated, invariably show massive habitat destruction. For

Figure 75. The Central Park Zoo. Located on six and a half acres off Fifth Avenue, between Sixty-Third and Sixty-Sixth Streets, the zoo is managed by the Wildlife Conservation Society and contains 130 species of animals in a variety of habitats. The zoo, which started as a small menagerie in the 1860s, was extensively modified in 1934 and then again in 1988. Photograph by Sidney Horenstein.

example, the gorilla group diorama includes the small cones of outlying volcanoes, which are forested. But the actual volcanoes are now converted to terraced farmlands.

Zoos do more than entertain and educate. They are often the last repositories for critically endangered species. One of the most famous captive-breeding programs is the one that saved the American bison, an effort undertaken by the American Bison Society at the Bronx Zoo and its supporters (not least of whom was Theodore Roosevelt) beginning in 1907. The fifteen bison that were rounded up at that time, housed at the zoo, and encouraged to breed were the progenitors of not only the herd currently on display at the Bronx Zoo but also a large number of the reconstituted herds still roaming in a wild, or semiwild, state in the American West. And of course modern-day genetics, when coupled with systematic efforts to conserve species though captive-breeding programs—such as the effort to conserve Asian tigers—pretty much ensures that species will not become extinct at least for the foreseeable future. But most of us, and most conservation NGOs, would prefer to see tigers and other iconic species persist indefinitely in the wild.

Another New York conservation initiative is the Center for Biodiversity and Conservation at the American Museum of Natural History. We noted in chapter 3 that the construction of the American Museum, beginning in the mid-nineteenth century, entailed the leveling of several small hills, the elimination of a small lake, and the deflection of a stream, typical first steps in the construction of a mammoth building complex in a major city. But if, in getting its start, the American Museum destroyed the existing terrain and local biota, it has more than repaid its debt to the world's remaining ecosystems and species diversity. Many of New York's doctors gained their interest in medicine from the various iterations of the Hall of Human Biology and Evolution. And the dioramas of landscapes, with their real specimens of fauna and flora from locales around the world, have been the first close-up experiences of the natural world for many a city kid. We ourselves were early visitors to the museum, and we ended up joining the scientific staff for our entire professional careers.

In a move that reflected our growing concern over the accelerating destruction of ecosystems and loss of species, we worked together (along with cocurator Joel Cracraft and many others, mainly from the museum's superb exhibition department) on the Hall of Biodiversity, which opened to critical acclaim in 1998 and was designated exhibit of the year in *Travel and Leisure* magazine. This was the first "issues hall" put together by the American Museum, meant to inform our visitors of the glories of the living world, as well as of the fact that virtually all of it faces dire threats.

The only true forerunner of this exhibition was the Grande Galerie at the Natural History Museum in Paris, which served in part as our inspiration. Viewing it, we were struck by the link it made between human activity and environmental disruption and species loss. Part of the exhibition featured a series of bas-relief minidioramas showing the woods along the Seine before the first permanent settlements were built, followed by scenes depicting changes in those woods during the growth of the modern city. The exhibit also featured a square meter of garbage encased in translucent plastic, said to be about one month's worth of trash discarded by a typical, small, apartment-dwelling Parisian family.

There are two basic sides to biodiversity: it is the evolutionary spectrum of life, and it is the ecosystems of the world, the arenas where the game of life is played out from moment to moment. You cannot understand life's diversity without knowing who the players are, that array of 30–40 million species, ranging from bacteria to redwood trees, from yeast cells to tigers. These are the evolutionary side of life's diversity. The combination of species endemic to a given part of the world, that are adapted to each other and the local climate, form the ecosystems of the world.

Thus we organized the Hall of Biodiversity by setting the evolutionary diversity of life across from our contemplation of the world's ecosystems. We built a "wall of life" with twenty-eight major divisions. This is still the only place in the world, we believe, where you can sample the entire gamut of life's diversity in one place: along a hundred-foot wall jammed with specimens and models of species culled from the museum's vast collections.

To show the ecosystems, we relied on modern technology such as slides and film, in a linear array, to portray the variation in the world's habitats. But we also drew on the museum's history of dioramas, building a walk-in experience of an African tropical rain forest. Museum scientists and exhibition staff, led by Joel Cracraft, traveled to the Dzanga-Sangha forest of the Central African Republic, where they collected and photographed specimens that became the basis for much of our New York City re-creation. We relied on the visual beauty of our specimens and re-created ecosystem to hook our visitors, but we also made the grim story clear: all of this is threatened. And we pointed to the culprit: human beings, whose numbers are rapidly increasing the world over.

Figure 76. A walk-through forest diorama in the Hall of Biodiversity at the American Museum of Natural History. The scene is modeled after the large and diverse Dzanga-Sangha rain forest of Central African Republic. © American Museum of Natural History/Enrico Ferrorelli.

Once we saw a ten-year-old run around a corner and into the main section of the hall, between the wall of life and the rain forest diorama. He skidded to a stop in his sneakers and, looking up at the sharks and other marine life swinging from the ceiling, said in awe, "Oh, wow!" So we know that at least some of our visitors get the beauty, the pure spectacle of nature. Do they get the message, though? Kids come in with their classes and fill out worksheets prepared in advance by their teachers, and our hopeful supposition is that at least some of them do get it. Follow-up surveys after the hall's opening seemed to suggest they do. The hall is kept up-to-date by "bio-bulletins," thoughtful video mini-essays on environmental themes that are posted several times a year on one of the screens in the exhibition.

The museum founded its Center for Biodiversity and Conservation at about the same time that we created the Hall of Biodiversity. Though the museum is the largest natural history museum in the world in terms of the square feet devoted to public exhibitions, it is also a thriving scientific research institution. Throughout its history, there have always been approximately forty members at a time on the curatorial research staff, whose job it is to amass collections and to publish research they do based on those collections. Margaret Mead was one of those curators, as were George Gaylord Simpson and Ernst Mayr. It has been a serious scientific center since its inception. Most of the research is biological in nature, and most of that biological research has centered on systematics: the discovery, description, naming, and classification of the living (and fossil) species of the world. Systematics is part of evolutionary biology and, along with ecology, lies at the heart of the study of the world's living diversity.

The museum has been affiliated with universities at least since the 1890s, when Henry Fairfield Osborn, a vertebrate paleontologist, was simultaneously professor of zoology at Columbia University and a curator of paleontology (and later president of the museum). Since those days, the museum has networked with the City University of New York, Yale, Rutgers, Cornell, and others. Many students got their PhDs while apprenticing at the museum. And in the first decade of the twenty-first century, the museum began its own graduate school program. Students affiliated with the museum's Center for Biodiversity and Conservation are enrolled in these programs and study with current members of the curatorial staff. The center is especially keen to host and mentor students from third-world nations, where many of the remaining diverse life-forms survive and are, of course, deeply threatened. This is one dramatic way that a New York institution is actively redressing the damage done and still to come: most of those students return home to take jobs of responsibility related to conserving biodiversity after their training in New York is complete.

The center, like the Wildlife Conservation Society in the Bronx, also has active programs that set up, monitor, and maintain conservation programs in critical areas around the world. The academic side of conservation depends on accurate assessments, including detailed inventories of what species actually exist in a given place, in order to develop and maintain successful conservation programs. This, of course, dovetails with the research interests of many of the curatorial staff. The museum has been a major participant in the Tree of Life program funded by the National Science Foundation, which aims to inventory all of the world's living biota and to understand the evolutionary history of all the living species of the world. This multifaceted conservation effort is largely funded by the United States government.

One of the more important programs conducted from afar by the Center for Biodiversity and Conservation is located in Vietnam, which harbors an astonishing variety of wildlife that somehow managed to survive decades of intense warfare. (Much like in the no-man's-land between East and West Germany, which was home to many species of European plants and animals despite the presence of land mines.) Once relations had normalized sufficiently to allow Western scientists into Vietnam, even species of large mammals were recorded there for the first time. Since then, the center's research staff and allied scientists have made many more discoveries while inventorying the biodiversity of Vietnam.

But the scientific work in Vietnam, conducted under the aegis of the Center for Biodiversity and Conservation, is not all that stands out. The center seems to have been largely successful in translating the science into policy—and advising the Vietnamese government on implementation of the country's plans to increase the area of lands included in its National Protected Area System. They have also identified routes used by poachers and illegal loggers. Even more arresting has been their success in working with local conservation officials to implement policy. As we shall see, it is local people who are most capable—more than external funding and the presence of Western biologists—of ensuring that protected areas are managed properly and effectively, so that wildlife protection occurs in the context of the culture and economic concerns of the people living in and adjacent to the protected areas. This is probably the ultimate challenge: the message of the world's international cities being translated to good effect to particular locales, connecting directly with foreign governments and trained professionals, but especially with the trained local people on whose shoulders it falls to translate data and policy into practical procedures— again, in concert with the traditions and economic needs of the local populace.

The museum's Center for Biodiversity and Conservation is also notable for its well-honed ability to network with other institutions. In Vietnam, the center's staff work with national, regional, and local governments. They also work with other NGOs, such as the World Wildlife Fund, and with scientists from other institutions in New York, around the United States, and elsewhere in the world, including, of course, Vietnam. Foundational support has also been an important factor.

The American Museum is an interesting hybrid, part public and in large measure private. The City of New York owns the original eighteen-acre parkland, as well as the twenty-three buildings that constitute the museum complex. The museum's board of trustees owns the fabulous collections housed there, collections of untold value amassed since 1869. In the old days, the trustees reportedly "passed the hat" to make up any spending deficits, especially when the storied banker J. P. Morgan was chairman of the board. Today, private support, beginning with the trustees themselves, remains a major source of funding. But so does financial input from the City of New York, whose mayor and several other officials serve ex officio as members of the board. Admission fees and other revenues are important, as are the grants the museum continually seeks and often wins from government agencies, such as the National Science Foundation and private foundations.

The museum is intensely networked with other scientific and cultural institutions, including, as already mentioned, universities. It is deeply involved in local, state, and national politics: President Barack Obama was at the museum in early 2011, lending his presence to a New

York City schools science fair. To many, the American Museum is a symbol of science in New York as important as research institutions such as Rockefeller University.

Finally, the media, especially the *New York Times,* pay close attention to the science and exhibitions of the American Museum. They help inform the public about the importance of biodiversity and about what is being done at the museum and elsewhere to figure out how to save it.

Thus the museum itself is a microcosm—a research and exhibition institution that is itself a single component of a network of varied institutions that share a common goal: to formulate ways to understand what biodiversity is and how it developed in the first place; the causes of the rapid loss of biodiversity (as the human population grows, and as habitats and their plant, animal, fungal, and microbial species consequently lose ground); and what can be done about it. The network is global, but virtually all the components are in cities. The next thing to consider is how policies derived from science and from the good intentions of city slickers in general can be—and are being—effected in the real world.

BOOTS ON THE GROUND: CONNECTING CITY-GENERATED WEALTH, SCIENCE, AND POLICY WITH CONSERVATION PROGRAMS AROUND THE WORLD

How do the institutions in cities, with all their expertise and financial clout, communicate their message and see it effected in areas around the world that both science and common sense have determined are the most critically endangered and otherwise important habitats and species? Working with national and regional governments, plus native NGOs, universities, museums, and zoos, can be difficult. Moreover, some of the larger American-based NGOs have acted more as rivals to each other than as cooperating partners in target conservation areas, even though the goals and motives of those NGOs are essentially identical. Still more difficult are the times when a program like the

American Museum's Center for Biodiversity and Conservation teams up not only with other American scientific and funding institutions but also with scientists from elsewhere in the world and, critically, native scientists, conservationists, and governmental agencies. Yet successful collaborations among diverse arrays of institutions and disciplines have become the norm, rather the exception, as this new international cooperative model for addressing the world's biodiversity crisis has matured over the past several decades.

Scientific consortia are adept at communicating with foreign national and regional governments, and they provide the fauna and flora surveys used to help identify areas of critical importance and potential success. Rarity of endemic species, especially iconic ones like Asian tigers, is one big criterion for determining where the work is most urgently needed. But of course, the work really entails finding and demarcating potentially conservable chunks of real estate—habitats— that simultaneously conserve ecosystems and their included species. Increasingly, too, the search is widened to identify potential wildlife corridors connecting a core area with others in the region, which enlarges the extent of a conservation area: the greater the amount of territory preserved, the more likely that species in it will be kept alive.

So the first phase of putting "boots on the ground"—that is, when working teams from cities gather in endangered habitats the world over—is, if not easily done, at least well established as a modus operandi of the conservation movement in general. But the final step is the hardest: implementing the programs on-site. Especially tricky is communication with the people living alongside wildlife in and around the designated conservation area. More often than not, they gather wood from the area for their cooking and heating fires, and for their building needs. They harvest fodder for their livestock and food for themselves, commonly hunting local animals and fishing in local waters.

No people on earth remain "pure" hunter-gatherers, living in small, often nomadic bands within local ecosystems, dependent solely on the productivity of their surrounding environment. But just as city dwellers eat freshly caught fish and buy furniture fashioned from woods growing wild in the forests of the world, which makes them still dependent on the productivity of wild places, so do millions of the world's people rely heavily on the resources available to them in the very places that conservationists would love to conserve. And this is a recipe for collision between the interests of organizations and their supporters—who often live half a world away—and the interests of the local citizenry, who almost invariably are told to either move out or, at the very least, stay out of the conservation area and keep their hands off the natural products they typically heavily rely on.

Most of the latter people are poor—desperately poor by the standards of the Western world. They have little income and subsist on the equivalent of approximately a dollar a day. They still need the now-protected forests, grasslands, and swamps they have traditionally relied upon. But outsiders now tell them they can no longer continue their tried and true ways of making a living, and all too often there are insufficient alternatives. Some conservationists simply hope that the people living in and around conservation-project areas will either die off or move away. Yet people will not simply go away for the sake of someone else's dream of conserving wildlife. Just as no one wants the garbage pits or toxic waste dumps near *their* houses, those of us in and around cities who send our dollars to well-meaning NGOs to help fund the preservation of living nature know or care little about the negative impact such actions often have on the lives of local inhabitants.

Humanitarianism is not the only motive for ensuring that the needs and interests of local people are looked after as they are forced to change their old ways of life. The more sophisticated policy makers at NGOs, universities, museums, and zoos engaged in hard-core conservation programs are deeply aware that without the active cooperation of local peoples—as administrators of, guides to, and guards of the conserved areas—the chances of long-term success in any given project are grim. Some policy makers have gone so far as to establish schools, providing a general education (reading and writing in English, but also in the local language, if necessary) and training local peoples in the elements of local biology and in the aims and techniques of conservation, so that some can assume active, paid roles in collaborating with the conservation project at hand. Water management and alternatives to slash-and-burn and other crude and harmful farming techniques are also increasingly a part of the training provided to local peoples.

To paraphrase the website for Duke University's Primate Center—which maintains active, consortium-based conservation projects in Madagascar—you cannot expect to conserve a critical area unless you simultaneously tackle the crushing poverty of the people living there. Indeed, the first local school we encountered that had been built by conservationists in response to their recognition that local people must be included in order for a project to succeed was at Ranomafana, in Madagascar, before it became a national park. That school had been established by the Duke Primate Center, especially by anthropologist Pat Wright (who is now at the State University of New York at Stony Brook), as part of the effort to conserve the forest at Ranomafana. And it was she who persuaded the Malagasy government to turn Ranomafana, in 1992, into Madagascar's third national park.

In a nutshell, local people must share in the influx of money if they are expected to cooperate. Some of that money comes from conservation NGOs and governmental sources involved, and much of it comes from ecotourism, from the increasingly large numbers of people who visit national parks and other conservation areas every year. Ecotourism is not always a good thing. It consists of city-based companies, most of them American or European, taking groups of primarily city people into wild areas. And although the stated goal is to have no negative environmental impact there, ecotourists are often unsuccessful in this.

Consider as an example the impact that tourism, as well as still-not-fully-regulated fishing, has had in the Galápagos Islands, the "Islas Encantadas" that played a critical role in the development of the evolutionary ideas of Darwin when he visited there in 1835. Now Ecuador's most affluent state, the Galápagos have been beset by tourists and by native Ecuadorians emigrating here in increasing numbers. The downside of "ecotourism" is there for all to see: despite the best intentions of ecotourists and the operators of ecotours, ships plying the routes between islands pollute the seaways close to shore. And the commercial traffic that supplies the tourist ships not only pollutes the waterways but also contributes directly to the tourism-based economy. The latter creates jobs but also expands the population of locals eager to work in the tourism trade, which in turn contributes to the environmental degradation now engulfing the Galápagos. The Galápagos are looking increasingly tawdry—and this, at least in part, is a result of the ability of so many of us urbanites to go to this fabled land.

Too often, too, the companies that run the programs to these often remote, hard-to-get-to places leave little money in the communities they directly impact. But on the bright side, ecotourism does indeed directly create jobs—for the aforementioned guides, guards, and administrators, who most often come from the ranks of local inhabitants. In Peru's Machu Picchu (a wonderful natural setting as well as an amazing archeological site), for example, only local Quechua Indians—descendants of the Incas—are allowed to serve as guides.

Time now to take a closer look at how cities speak to the local inhabitants—and how the locals speak back, ultimately, one hopes, in service of successful conservation and a stabilization of the Sixth Extinction.

CORBETT NATIONAL PARK AND PROJECT TIGER: INDIA

Large mammals are at the top of the list of species that people most love and want to see survive. On safari in southern Africa, everybody wants to see the big four: lions, elephants, giraffes, and leopards. Depending on what part of the continent people visit, white rhinos, black rhinos, and hippos may also make the list. All these animals are part of the "charismatic megafauna" and, along with koalas, pandas, and certain other mammalian species, represent the conservation movement. (We were jolted one day to see the World Wildlife Fund logo, featuring a panda, on the side of a car in downtown Antananarivo, Madagascar, in the late 1980s.)

But for many people the world over, the Asian tiger stands out as most iconic. Beautiful, fearsome, secretive, and by all accounts highly endangered, tigers seem especially to resonate with city dwellers, who do not have to live alongside them. The latest census of the Indian tiger population (released in April 2011) estimates that there are 1,700 tigers currently living in India—up from the 2006 estimate of 1,411. This was good news to the people of the Indian conservation organization Project Tiger and others concerned with the survival of the Bengal tiger. But skeptics were quick to point out that thirteen areas had been added to the location covered by the latest census, and those new areas accounted for all but 7 of the additional number of tigers registered by the new census. Indian tigers are barely holding their own.

A century ago there were an estimated 100,000 tigers in India. The famed hunter (*shikari*) Jim Corbett, whose mid-twentieth-century books were written in his later years, saw it coming. Corbett grew up in the Himalayan town of Nainital, where the family spent their summers within sight of the high peaks once thought to be perpetually covered in snow and ice. They wintered in the area where the Shivalik Hills (Himalayan foothills) give way to the Indian plains, a landscape covered by dense forests and woodlands, in which game was abundant and leopards and tigers always present. Corbett shot his first leopard when he was nine.

Corbett's books center on his remarkable success in ridding hill villages of the scourge of man-eating leopards and tigers. He shot his first tiger in 1907, and shot many more in the ensuing decades. But by the 1920s, seeing the handwriting on the wall, Corbett increasingly laid his rifle aside in favor of his cameras, and he took probably the earliest motion pictures of tigers in the wild. He could see that deforestation (some of which he had directed while working for the railroad as a young man) was rapidly diminishing habitat and, therefore, the numbers of tigers and all other fauna.

Other "sportsmen" like himself contributed to the depredation, some of them taking part in big, organized events, with beaters to drive the animals, and politicians and rajas riding in howdahs on the backs of elephants, their guns at the ready. "Poaching" by the villagers themselves also figured in the demise of wildlife at a rate faster than reproduction, and migration from neighboring Nepal, could replenish. Corbett saw all that, and he began to write and speak of the need to conserve.

It was apparently the result of some idyllic fishing and hunting adventures with Corbett along the Ramganga River, west of Nainital and Kaladhungi, that Malcolm Hailey, the British governor of the Indian state of Uttar Pradesh, proposed designating the area as a national park. With Corbett's help in defining the boundaries, Hailey National Park became India's first national park, in 1936. After a short postindependence interval, when the park was renamed Ramganga National Park, it was then officially renamed Corbett National Park.

Thus the impetus for the park and the conservation of Indian wildlife came from a hunter-turned-conservationist who was essentially a native Indian, and whose hunting and fishing companion also happened to be the governor. Governmental authority is always necessary in setting aside land and budgeting funds to maintain a protected area. True, Corbett was of British descent, but his family had been there for a long time before his birth. The Corbetts were "domiciled," meaning they were not of the same upper crust as the British politicians and businessmen. Corbett spoke many dialects of Hindi, and he once said that he preferred the superstitions of the superficially simple Indian hill folk to those inculcated in British schoolchildren.

The next significant step came from the New Delhi government. In 1973, Prime Minister Indira Gandhi secured protected status for the Bengal tiger, and helped form Project Tiger, as part of the Wildlife Protection Act of 1972. Project Tiger has since been reorganized and is now run by the National Tiger Conservation Authority, established in 2005. It is administered by the inspector general of forests—and so remains firmly in the hands of the Indian national government. But the project counts on board membership by politicians, by local tribal peoples, and by scientists the world over. The World Wildlife Fund, headquartered in Washington, D.C., has been a significant donor, as have other NGOs.

So far, the story of Corbett National Park, the first and most important of India's tiger reserves under Project Tiger, is comparable to conservation ventures around the world. But the collision between national and international interests, on the one hand, and the interests of the poor people living in and alongside Corbett National Park, on the other, has a dimension not found in any other conservation effort in the world. The local people are still being attacked and eaten by tigers—the poster species that prompted the effort to conserve ecosystems and species threatened with extinction.

Corbett had thought that the man-eaters he killed (and others he had only heard about) were anomalies, that humans were not a regular item in the tiger diet. It is true that as humans spread around the globe over the past forty thousand years, they were already experienced and savvy hunters, and they easily pushed many large species to the brink—and over the edge—of extinction, including large carnivores. The large mammals of Europe and Asia, and perhaps especially of North America (where nothing like a human being had ever appeared before), simply did not know how to cope with this new predator that suddenly appeared in their midst. But tiger attacks on humans in India are not all that rare. In late 2010 and early 2011, six people from Sundarkhal Village, near the eastern margin of Corbett National Park, were killed by what turns out to have been a pair of tigers. Since then, other tiger attacks on humans have been reported in the vicinity.

This constitutes a serious collision. Many people in Delhi, Mumbai, and Kolkata, like people in London, Paris, Los Angeles, and New York, simply love tigers, but they don't have to live beside them. They don't live in huts without electricity or running water. They don't have to cross the two-lane highway to gather fodder for their livestock or leaves from curry trees for their own cooking pots. The six to ten tigers living on the eastern edge of Corbett National Park regularly cross that same highway, near Sundarkhal Village, to drink from the Kosi River, the only local water source for both people and wildlife during the dry season. Tigers are protected, and their numbers are growing, however slowly, in Corbett. There is plenty for them to eat: the sambar and especially chital deer are abundant. But as tiger numbers grow, and as the human numbers do the same, meetings are increasingly likely. And likely to prove deadly.

Project Tiger promotes the idea of relocating villagers—which it correctly says must be done with sensitivity. The advantages to the villagers are said to be improved amenities and enhanced access to schooling for the children. The advantage to the tigers is that, with the opportunity to eat humans minimized, they will inadvertently minimize the sort of bad press they have been getting of late.

But consider the reaction to the tiger attacks outside Corbett. They made headlines in India and, via the Internet, around the world. Conservationists tended to be outspoken, defending not tigers as a species, which they are pledged to defend, but the particular tiger (it was at first thought the attacker was a single individual) who attacked and ate the human victims. Nor was it only Indian citizens who defended the tiger. Discuss this story in New York, as we have, and you will find no dearth of people who seem not a bit self-conscious or embarrassed when they say there are plenty of people in the world and not enough tigers. In fact, suspicion that the attacks were somehow actively provoked by the victims was rife—and duly reported in the media as possibly true.

In those media reports, it was difficult to find much said on behalf of the people of Sundarkhal whose relatives had been the victims. Poignantly and eloquently, a Sundarkhal villager asked, "Are we not people too?" The government for its part temporized. After all, the purpose of the park was especially the conservation of tigers, and to kill one seemed antithetical. Park staff tried using a large tiger trap kept in the vicinity, with the aim of trucking the animal to some remote region. (Conservationists involved with tigers hate the idea of sending miscreant tigers to zoos, although many have ended up consigned to zoos. They can at least be used for breeding purposes, and it's a better fate than their being killed outright.) But the trapping attempt didn't work.

Poor people the world over have little political clout. But after the sixth kill (perhaps significantly, the sixth victim was a young man; the previous ones had all been women), the terrorized neighbors, who lived only a few miles up the road from posh hotels housing ecotourists, took to the street and shut it down. There are few roads in that region of India, so not only the comings and goings of tourists to Corbett National Park were disrupted but so was the normal commerce between towns.

It worked: the next day, the government authorized the shooting of the tiger, which was found lying near its kill, the young man whose body it had dragged down to the edge of the Kosi River. Mustering six tamed elephants, hunters shot the tiger the old-fashioned away—from a safe perch. The tiger lay quiet while the elephants approached. Human remains were reported to have been found in its stomach. About a month and a half later, another tiger began killing Sundarkhal villagers. This time the people took to the street right away, marching down the road and preventing the officials and board members of Project Tiger—who had gathered there for a long-planned meeting, and who no doubt were debating the significance of the just-announced results of the latest tiger census—from going to their next meeting. This time the trapping attempt worked. The current whereabouts of the tiger are, however, unknown.

One bright light in this story is the small but vital cadre of local park rangers, foresters, guides, and guards involved with Corbett National Park and tiger conservation. They are Kumaonis, who speak the local Hindi dialect and observe local customs and eat local foods. They also speak perfectly good English and are excellent naturalists who have been schooled right there in the region. They know how to use cameras, binoculars, and the Internet far better than most of the ecotourists they take into the park everyday. As locals, they share the concerns of other locals, including their horror at the attacks. They are the ones whom people in cities must continue to speak to and support, and to consult on what needs to be done next to foster the integration of the

money and savvy of those in the cities with the pressing need to conserve the world's wildlife. You can't expect people to conserve something that is eating them. But in educated young people, wise to the
ways of both nature and city-bred conservation, there is hope for a better connection to meet the interests of all concerned.

BACK TO THE CITIES: POLICY AND POPULATION

There's no doubt about it: human overpopulation, and the consequent
collisions between people and the world's ecosystems and species, is the
root cause of the present-day Sixth Extinction. It began when hominids
left Africa and spread throughout the Old World. But the first big blow
came when our own species, *Homo sapiens,* began leaving Africa in significant numbers about forty thousand to fifty thousand years ago. By
about two thousand years ago (the time when Madagascar apparently
was first colonized), our species had reached every habitable point on
the planet. Our arrival everywhere was quickly followed by major
extinction events, notably among the larger species of mammals.

But that was just the beginning. Agriculture, with its necessary
implication of a settled existence, began in earnest in the Old World
about ten thousand years ago, when the earth is generally thought to
have housed 5 million human beings. As we have seen, taking over our
own food production soon began to produce humans at a rate much
faster than they were dying off—despite the persistent threat of famine
related to failed rainfall and other natural disasters. The explosive
growth of the human population is the main underlying cause of the
second wave of the Sixth Extinction, currently under way.

Our population growth rate over the past ten thousand years has
been logarithmical: that is, it started slowly but gained momentum. A
billion people lived on this planet around 1800; the present number of
more than 7 billion, reached in 2011, represents more than a doubling of
the global population within our own (that is, the authors') lifetimes.
And although most reports maintain that the *rate* of human population

growth has been slowing since a peak of 2.2 percent in the early 1960s, fast rates of growth are still the norm in many, especially poor, nations around the world. Half the world's population lives in poverty, and growth rates still project a total of over 9 billion people by 2050.

Human pressure on the world's ecosystems and species is only going to worsen if our population is left unchecked. And part of the answer to limiting the extent of the Sixth Extinction lies in addressing the means to further stabilize human population growth. And that means policies, planning, and financial support emanating from governmental agencies and NGOs—which are located, for the most part, in cities. One initiative designed to limit population growth is the well-known one-child-per-family law governing primarily the Han populace in China, which comes directly from the central government in Beijing. But most promising are the initiatives that seek to slow population growth by encouraging economic development (where possible). In particular, the inclusion of women in educational programs enables them to enter their local workforces in greater numbers, serving the dual purpose of improving lives and reducing birthrates.

But while some governments, such as those of China and India, are concerned about population issues and actively seeking to promote stabilization of their own numbers, governmental organizations in general are slow to advocate population policies. This is especially clear in the case of the United Nations, which is the best source of data on rates of population growth and urbanization (the history of population growth in the world's largest cities). But the United Nations has made it clear in UN publications that it makes no judgments on the nature or significance of the numbers and estimates it publishes. This includes projections of world population growth as a factor of known and estimated fertility rates, as well as reports of where growth remains highest (in thirty-nine countries in Africa; in nine in Asia, in six in Oceania, and in four in Latin America, according to a May 3, 2011, press release by the UN).

The reason the UN avoids advocating population policies is simple: Population issues, especially initiatives that seek to slow population growth, are inherently controversial. The political right in the United States and elsewhere, especially the religious right, opposes family planning, including the dissemination of birth control medications and devices and instructions on their use. And the right abhors abortion. Indeed, the Catholic Family and Human Rights Institute in New York issued an essay titled "The Ford Foundation: Founder of Modern Population Control," an interesting read that addresses the history of the Ford Foundation's funding of population policies and programs. In this essay, the institute adopts the strategy of assigning guilt by association, claiming that the Ford Foundation's involvement in population policies and programs at all is proof that it has long been up to no good.

But opposition to population policies that are aimed at stabilizing population growth, whatever the means advocated and funded, comes from the left as well. The left has always condemned anything that smacks of eugenics, the nineteenth-century movement that advocated selective breeding to enhance the genetic health of, and improve, the human stock—which is much like what farmers have done for at least ten thousand years, and a process that genetic engineering seeks to perfect in modern times (again, much to the dismay of some members of the political left). Eugenics has been explicitly involved with programs to eradicate genetic disease, but also with programs designed to curtail the reproductive lives of the poor and disenfranchised in the United States and the world over. Any attempts to curb the rights of the poor, including their right to have as many children as they want, is bound to inflame many on the political left. So, in the issues of population regulation and birth control, governments usually have their hands tied, unless they happen to be in total control in a nondemocratic setting. That leaves mainly NGOs, such as the Ford Foundation, Planned Parenthood, and others.

Consider, briefly, the New York–based Ford Foundation. Perhaps because of political pressures, actual and potential, the foundation currently makes no mention of world population in the priorities it lists on its website. What it does promote is a commitment to sexual and reproductive health and rights, to sustainable development, and generally to human rights and social justice the world over, as well as to other worthy initiatives. But given that population growth, and the resulting pressures on ecosystems and species, is the root cause of the Sixth Extinction, remedies must go beyond population *stabilization*. These remedies must embrace especially the human rights of those who live in poor regions, where most of these problems exist and are especially intense, but also promote sustainable, improved economies for these people.

Cities are critically important to the development, promulgation, and implementation of such programs. Despite the inevitable, predictable political opposition such initiatives face, their programs are not at base political or politically motivated. They are well-thought-out examples of common sense. NGOs like the Ford Foundation, which rely on endowments and private contributions to pursue their objectives, are vital to the efforts to ameliorate the many interconnected global problems the world now faces, including the Sixth Extinction.

CONCLUSION

Cities are the best of civilization—and often the worst of civilization. They are, of course, here to stay, and the future promises even greater concentrations of humanity in cities. Futuristic scenarios, as well as concrete plans emanating from urban studies, architecture, and the social sciences, suggest ways of building even denser concentrations of affordable housing, thereby reducing sprawl and freeing up space for more parks and perhaps relieving the countryside of some of its human population. There have been exoduses from the countryside to the cities in the past, as happened in Brazil in the mid- to late twentieth century, and a bit earlier in the United States, when the migration of African Americans from the Deep South to Chicago, New York, and other cities, took place. But the full answer to global population pressure on the environment is unlikely to be found in the further concentration of humanity in cities.

Cities are resilient, and the process of cleanup and restoration continues in cities such as New York, Beijing, and Delhi, among others. These efforts depend on money and willpower; and as we have seen, there are encouraging signs of restoration and renewal as some of the native flora and fauna reappear in cities.

But the point of cities is not to return them to a naturally wild state. That is impossible and not even particularly desirable. People seem ever-better adapted, culturally, to urban life. We are, after all, a self-absorbed species. People in cities tend not to know where their food, water, and electricity come from, at least not in specific detail. Even if they do know about such things, they really do not, and need not, care. But they must care about biodiversity, and even must redeem themselves by reaching out to regions where biodiversity can still be restored and maintained. Cities can help supply the necessary knowledge and financial resources that will permit people in other regions to do more than simply watch (and promote) the disappearance of their wildlife. Cities can help give them a better chance of conserving their own natural heritage. Many people, especially in scientific research organizations and NGOs founded for that very purpose, do perceive the need for such involvement. Part of their outreach is to educate others, both locally and internationally, about the reasons why a healthy global ecosystem is essential to the long-term healthy survival of the people of the world, city dwellers and country folk alike.

We continue to witness the havoc caused by changed weather patterns resulting from increased carbon dioxide emissions. We see that the accelerated warming of the planet is melting glaciers, raising the sea level, and threatening to flood the North Atlantic with freshwater that will inevitably kill off the oceanic food chain there and ruin the fisheries ... and on and on. It is we in the cities who must see that our own existence—every bit as much as the existence of everyone else on earth—depends upon our direct outreach. Only we can save the ecosystems and species still clinging to life here.

NOTES, REFERENCES, AND SUGGESTIONS FOR FURTHER READING

Much of this narrative is based on our own experiences and observations—of New York City, first and foremost, but also of the majority of the places outside New York the world over that are singled out in the text. We've both had the opportunity to travel extensively—sometimes together, but most often not. We gratefully acknowledge—and cite—the opportunities we've both had for becoming familiar with environmental issues (and other cities!) as lecturers in the American Museum of Natural History's "Discovery Tours" ecotourism program since the late 1970s. When we talk about places like Cairo, Delhi, Mexico City, Cape Town, and Rio (and so many more)—we speak with the background of people who have been there, and who have done enough research to speak about these places authoritatively.

That said, of course, we have relied extensively for this book on published sources—and we are aware of other resources that could serve very well as additional reading on many historical and modern environmental points—both local (New York City) and global. When we first discussed writing this book, our thought was that governments—who, after all, "own," and at the very least control, the lands of this earth—would be the key to the outreach of cities like New York in the goal of ameliorating the ongoing epidemic of ecosystem and species destruction. And this, of course, remains true: the interest and active involvement of governments everywhere (as in the case of Project Tiger and the conservation efforts at India's Corbett National Park, as well as elsewhere in India, discussed in chapter 8) are perhaps the key elements. But in recent years, especially as the global economy has weakened and sagged in the new millennium, the importance of nongovernmental agencies has, in a sense unexpectedly, become increasingly evident. And for information on the activities of these various agencies, you cannot do any better, at least as a start, than their often elaborately detailed websites. We realize that these websites are by their very nature self-serving, but the raw statistical data and overall nature of their conservation programs do, we think, get communicated credibly.

In recent years, the credibility and authenticity of resources posted on the Internet have become not only more detailed but also more reliable. When people identify an example of an Internet resource supposedly "not to be trusted," frequently they point to Wikipedia (and we hasten to say that none of the research in this book is based on Wikipedia articles!). But nonetheless, it is well worth acknowledging that, as we rechecked points and looked to see what is available for further reading, it turned out that Wikipedia seems to be a lot more trustworthy now, and its articles, with their bibliographies, are a great source for additional reading. Check out, for example, the "Newtown Creek" Wikipedia entry. Such articles constitute additional resources on topics ranging from local New York City issues and places to the international issues and places that we've mentioned in the text.

PREFACE

The three environmental books one of us (Niles Eldredge) published in the 1990s are *The Miner's Canary: Extinctions Past and Present* (New York: Prentice Hall Books, 1991); *Dominion* (New York: Henry Holt, 1995; paperback ed., Berkeley: University of California Press, 1997); *Life in the Balance: Humanity and the Biodiversity Crisis* (Princeton, NJ: Princeton University Press, 1998).

CHAPTER 1

For United States Census statistics, see www.census.gov. Much of the ecological theory underlying our discussion of how the invention of agriculture radically changed the ecology of our species, and many of our conclusions, is developed more fully in: Niles Eldredge, *Dominion* (New York: Henry Holt, 1995; paperback ed., Berkeley: University of California Press, 1997); and Eldredge, *Life in the Balance: Humanity and the Biodiversity Crisis* (Princeton, NJ: Princeton University Press, 1998).

For more information on the peopling of the Americas, see the special issue on the topic: Rolando González-José, ed., special issue, *Evolution: Educa-*

tion and Outreach 4, no. 2 (2011). On Brooklyn and agriculture, see Marc Linder and Lawrence S. Zacharias, *Of Cabbages and Kings County: Agriculture and the Formation of Modern Brooklyn* (Iowa City: University of Iowa Press, 1999).

Environmental studies on pollution are rapidly increasing and changing the picture for better and (all too often) for worse every year. The Internet, of course, is a handy source for the latest monitoring results on air, water, and soil conditions globally. Our brief discussions of some of the world's most polluted urban environments are meant to provide snapshots of conditions as they existed while we were writing. Not only do the data change but also Internet articles disappear. For example, the World Bank's 2002 study on air pollution in Cairo seems no longer to be available.

For recent articles and data on environmental problems in some of the large cities we discuss here, see the following. On São Paulo's environmental challenges, see www.gef.or.jp/20club/E/SaoPaulo_e.pdf; on New Delhi's environmental concerns, see http://Delhi.gov.in Home Economic Survey of Delhi Content; and for the famous City-of-the-Dead slum in Cairo, see http://blogs .newschool.edu/epsm/2012/05/24/city-of-the-dead-cairo. On housing density and sprawl in Los Angeles, see Blaine Harden, "Out West, a Paradox: Densely Packed Sprawl," *Washington Post*, August 11, 2005.

Interested readers will easily find more up-to-date information on the environmental conditions of all the world's major cities by searching on the Web for reputable governmental and NGO websites.

Our thanks to Professor Arthur M. Shapiro for bringing to our attention a multiauthored book with a scope and purpose similar to our own: Jose Athor, ed., *Buenos Aires: La historia de su paisaje natural* (Buenos Aires: Fundación de Historia Natural "Félix de Azara" y Universidad de Maimonides, 2012).

For an introduction to New York granite and marble buildings, see Sidney Horenstein, "Building Stone Treasure Troves," *Evolution: Outreach and Education* 1, no. 4 (2008): 520–30.

CHAPTER 2

For general resources on the history and natural history of New York, including precolonial and colonial New York, see Eric Sanderson and Markley Boyer, *Mannahatta: A Natural History of New York City* (New York: Harry N. Abrams, 2013); Kenneth Jackson et al., *The Encyclopedia of New York City*, 2nd ed. (New Haven, CT: Yale University Press, 2010); Peter Eisenstadt et al., *The Encyclopedia of New York State* (Syracuse, NY: Syracuse University Press, 2005).

The go-to resource for information on Central Park is S.C. Miller, *Central Park: An American Masterpiece* (New York: Harry N. Abrams, 2003). See also Robert McCabe, *The Ramble in Central Park: A Wilderness West of Fifth Avenue* (New York: Abbeville Press, 2011); Roy Rozenzweig and Elizabeth Blackmar, *The Park and the People: A History of Central Park* (Ithaca, NY: Cornell University Press, 1998).

For more information on Inwood Hill Park, see Sidney Horenstein, "Inwood Hill and Isham Parks: Geology, Geography, and History," *Transactions of the Linnaean Society of New York* 10 (2007): 1–54, and for aspects of postglacial geology, see Horenstein, "New York City Mastodons: Big Apple Tusks," *Evolution: Outreach and Education* 1, no. 2 (2008): 204–9. For the quote by Arthur Graves, see "Inwood Park, Manhattan," *Torreya* 30, no. 5 (September-October 1930): 117–29.

For more information on the geology of the New York City region, see Charles Baskerville, *Bedrock and Engineering Geologic Maps of Bronx County and Parts of New York and Queens Counties*, U.S. Geological Survey Miscellaneous Investigations Series (New York: U.S. Geological Survey, 1992), Map I-2003, scale 1:24,000; Baskerville, *Bedrock and Engineering Geologic Maps of New York County and Parts of Kings and Queens Counties, New York, and Parts of Bergen and Hudson Counties, New Jersey,* U.S. Geological Survey Miscellaneous Investigation Series (New York: U.S. Geological Survey, 1994), Map I-2306, scale 1:24,000; Pamela Chase Brock and Patrick W.G. Brock, "Bedrock Geology of New York City: More Than 600 M.Y. of Geologic History," Stony Brook State University of New York, Department of Geosciences website, 2001, www.geo .sunysb.edu/reports/ny-city/. Additional papers related to New York City geology can be found by searching the postings by Charles Mergurian.

For more details on the history and environmental impact and ramifications of the Panama Canal, see Niles Eldredge, *Life in the Balance: Humanity and the Biodiversity Crisis* (Princeton, NJ: Princeton University Press, 1998).

CHAPTER 3

It is, of course, unclear what the terms *ownership* and *sale* might have meant to whichever of the Native American groups negotiated with the Dutch. Native American economies of the time were an amalgam of hunting and gathering with agriculture. While no "pure" hunter-gatherer groups remain on earth, those who survived initial contact with European and other peoples long enough to be scrutinized (for example, the San people of southern Africa) had no sense that land was owned by individuals, although bands, of course, had their home ranges, their "territories." Agriculture, however communal it may be, necessarily creates greater ties to specific places, and so it is not impossible that Native Americans realized they were ceding the rights to Manhattan and other nearby places that they sold. See Robert S. Grumet et al., *Historic Contact: Indian People and Colonists in Today's Northeastern United States in the Sixteenth through Eighteenth Centuries* (Norman: University of Oklahoma Press, 1995).

An indispensable book on New York's history is E. G. Burrows and M. Wallace, *Gotham: A History of New York City to 1898* (New York: Oxford University Press, 1998).

For a pictorial history of the development of Manhattan and surrounding regions, see Eric Homberger, *The Historical Atlas of New York City* (New York: Henry Holt, 2005). The account of the purchase of Manhattan Island by the Dutch from the Native Americans for "$24 of junk jewelry" is in Stan Freberg, *Stan Freberg Presents the United States of America*, Rhino/Wea, 1996, audio CD. Homberger reports that New York exported a total of just over $3 million in commodities in 1803, a figure that grew to over $16 million a mere four years later in response to the efforts of John Jacob Astor and, specifically, his involvement in the fur trade.

For more on the history of public health in New York City, see John Duffy, *A History of Public Health in New York City, 1625–1866*, vol. 1 (New York: Russell Sage Foundation, 1968). For the building of the American Museum, we rely especially on an unsigned article on the history of the American Museum that was based on an exhibit prepared by Katherine Beneker and on view in 1958 and 1959. The museum's exhibit was developed to "acquaint its visitors with the changes that have taken place," since "over the years the face of the park has greatly changed, owing in large part to the growth and development of the Museum." This article was published originally in the museum's journal, *Curator.* Quotes by Albert Bickmore are from Katherine Beneker, "Theodore Roosevelt Park: 1807–1958," *Curator: The Museum Journal* 3, no. 2 (April 1960): 161–82, quotes on pp. 162, 166.

For information on Collect Pond, Peter Kalm, Hell Gate, and Battery Park, see the following. On Collect Pond, consult Kenneth Jackson et al., *The Encyclopedia of New York City*, 2nd ed. (New Haven, CT: Yale University Press, 2010); and Eric Sanderson and Markley Boyer, *Mannahatta: A Natural History of New York City* (New York: Harry N. Abrams, 2013). On Tea Water Pump, see Louis Pope Gratacap, *Geology of the City of New York (New York)* (New York: H. Holt,

1909), 21. On Peter Kalm, see Adolph B. Benson, *Peter Kalm's Travels in North America: The English Version of 1770*, vols. 1 and 2 (New York: Dover, 1987).

On Hell Gate, consult Jackson et al., *The Encyclopedia of New York City*. Captain Dermer's quote about Hell Gate is found in John Thomas Scharf, *History of Westchester County, New York* (Philadelphia: L. E. Preston, 1886), 1:40. On the Battery, see Rodman Gilder, *The Battery* (New York: Houghton Mifflin, 1936).

For the quote by Robert Moses on the Brooklyn-Battery Tunnel, see Robert Caro, *The Power Broker: Robert Moses and the Fall of New York City* (New York: Vintage, 1975), 641.

CHAPTER 4

Information on the Dakota and its surrounds has been drawn in part from an article by Christopher Gray, "Streetscapes/73rd Street between Columbus and Amsterdam Avenues: An 1880's West Side Block with Many Changes," *New York Times*, October 19, 2003; and an anonymous, undated article, "The Dakota," New York University, www.nyu.edu/classes/finearts/nyc/upper west/dakota.html, accessed January 5, 2011.

On New York City planning and growth, see Hillary Ballon, *The Greatest Grid: The Master Plan of Manhattan, 1811–2011* (New York: Columbia University Press, 2012); the quote about the commissioners' duties is on p. 30, and part of the Commissioners' Plan of 1811 is reproduced on p. 40. Also see Hillary Ballon and Kenneth T. Jackson, *Robert Moses and the Modern City: The Transformation of New York* (New York: Norton, 2007); John W. Reps, *The Making of Urban America: A History of City Planning in the United States* (Princeton, NJ: Princeton University Press, 1992); David Schuyler, *The New Urban Landscape: The Redefinition of City Form in Nineteenth Century America* (Baltimore: Johns Hopkins University Press, 1986); David M. Scobey, *Empire City: The Making and Meaning of the New York City Landscape* (Philadelphia: Temple University Press, 2003). The Randel quotes are from John Randel Jr., "City of New York, North of Canal Street, in 1808 to 1821," in *Manual of the Corporation of the City of New York*, ed. David T. Valentine and Samuel J. Willis (New York: E. Jones, 1864), 848, 850. The Bridges quotes come from William Bridges, *Map of the City of New-York and Island of Manhattan with Explanatory Remarks and References* (New York: T. and J. Swords, 1811). The quote by Clement Clarke Moore comes from Robert T. Augustyn and Paul E. Cohen, *Manhattan in Maps: 1556–1995* (New York: Rizzoli, 1997), 103.

On railroads and subways, see Eric Homberger, *Historical Atlas of New York City: A Visual Celebration of 400 Years of New York City's History* (New York: Henry

Holt, 2005), 76–77, 107. See especially the map Homberger provides of the earliest rail lines in New York, "New York Railroads," p. 77. We relied especially on J. W. Greene, "New York City's First Railroad: The New York and Harlem, 1832 to 1867," *New York Historical Society Quarterly Bulletin* 9 (January 1926): 107–23, for this section. For more on railroads, see also Peter Derrick, *Tunneling to the Future: The Story of the Great Subway Expansion That Saved New York* (New York: New York University Press, 2002); Carol Sheriff, *The Artificial River: The Erie Canal and the Paradox of Progress, 1817–1862* (New York: Hill and Wang, 1997). For the Reeve quote, see Arthur B. Reeve, "The Romance of Tunnel Building," *World's Work* 13 (December 1906): 8343.

CHAPTER 5

On garbage, see Benjamin Miller, *Fat of the Land: Garbage of New York: The Last Two Hundred Years* (New York: Basic Books, 2000); Anne Buttenwieser, *Manhattan Water-Bound: Manhattan's Waterfront from the Seventeenth Century to the Present* (Syracuse, NY: Syracuse University Press, 1999); Kevin Bone, ed., *Water-Works: The Architecture and Engineering of the New York Water Supply* (New York: Monacelli Press, 2006); Gerard T. Koeppel, *Water for Gotham: A History* (Princeton, NJ: Princeton University Press, 2000).

On Newtown Creek, the best resources are various entries at the U.S. Environmental Protection Agency website, www.epa.gov. For the Department of Environmental Protection's quote about Newtown Creek sewage treatment, see "Newtown Creek Sewage Treatment Facility," at the Habitat Map website, n.d., http://habitatmap.org/markers?marker_id=144-newtown-creek-sewage-treatment-facility, accessed March 27, 2014.

CHAPTER 6

On alien (invasive) species in New York, see New York State Department of Environmental Conservation and New York State Department of Agriculture and Markets, *Final Report of the New York State Invasive Species Task Force* (n.p.: New York State Department of Environmental Conservation and New York State Department of Agriculture and Markets, fall 2005), www.nyis.info/pdf/NYS_ISTF_Final_Report.pdf.

On federally protected plants in New York State, see "Federally Protected Plants," New York State Department of Environmental Conservation, n.d., www.dec.ny.gov/animals/7133.html, accessed March 12, 2011; and "New York

Natural Heritage Program," New York State Department of Environmental Conservation, n.d., www.dec.ny.gov/animals/29338.html, accessed June 1, 2011. The following statement appears on the latter web page: "Our mission is to *facilitate preservation of New York's biodiversity* by providing comprehensive information and scientific expertise on rare species and natural ecosystems to resource managers and other conservation partners" (emphasis added).

On Steve Brill, see his website Wild Food!, www.wildmanstevebrill.com.

For the latest information on Dutch elm disease, chestnut blight, the ash borer, the Asian longhorned beetle, and other ongoing problems caused by invasive species, see the U.S. Department of Agriculture National Agricultural Library website, "National Invasive Species Information Center: Gateway to Invasive Species Information," last modified February 27, 2014, www.invasivespeciesinfo.gov. On bedbugs, see "12 High-Profile NYC Bedbug Scares," *The Week,* October 28, 2010, http://theweek.com/article/index/206173 /12-high-profile-nyc-bedbug-scares.

On the section "What's in the Water," and especially for information about the Gowanus Canal, see news items on the Riverkeeper website, www.river keeper.org, and "Gowanus Canal History," Gowanus Dredgers Canoe Club website, n.d., www.gowanuscanal.org/history.html.

For Washington's quote in the sidebar, see *Diary of Washington: From the First Day of October, 1789, to the Tenth Day of March, 1790* (New York: n.p., 1858), 60.

CHAPTER 7

On New York City parks, see Alice Deutsch, ed., "Natural History of New York City's Parks and Great Gull Island," *Transactions of the Linnaean Society of New York* 10 (September 2007). For the quote about the promise of a new Fresh Kills park, see Linda W. Foderaro, "From Dump to Paragon of Ecology: A First Look," *New York Times,* September 30, 2011. For the policy statement on the restitution of Inwood Hill Park, see Tanja Crk, *Inwood Hill Forest Restoration,* Columbia University website, last updated December 20, 2006, www.columbia.edu /itc/cerc/danoff-burg/RestoringNYC/RestoringNYC_InwoodForests.html.

For further information on Alley Pond Park, Flushing Meadows, the High Line, Fresh Kills, and Riverbank State Park, see the websites of the New York City Department of Parks and Recreation, www.nycgovparks.org, and the New York State Department of Environmental Conservation, www.dec .ny.gov.

The story of the Protectors of Pine Oak Woods is told by Bruce Kershner, *Secret Places of Staten Island* (Dubuque, IA: Kendall/Hunt, 1998).

The book on Robert Moses mentioned in this chapter is Robert Caro, *The Power Broker: Robert Moses and the Fall of New York City* (New York: Vintage, 1975).

For Bart Chezar's oyster story, see Ben Muessig, "Shell Shocker! Oysters Thrive in Our Dirty Water," *Brooklyn Paper,* November 13, 2008.

On the plan to reduce nitrogen discharge by 50 percent by 2020, see "Excess Nitrogen in the Water of Jamaica Bay," New York City Audubon Society, n.d., http://nycaudubon.org/issues-of-concern/excess-nitrogen-in-the-water-at-jamaica-bay, accessed March 27, 2014.

For the Park Department's remarks about the future of Freshkills Park, see "Freshkills Park," New York City Parks website, n.d., www.nycgovparks.org/park-features/freshkills-park, accessed April 5, 2014.

On the Department of Environmental Conservation's remarks about the cleaner condition of the Hudson River today, see "Cleaner Beaches on the Hudson River," part of the *Hudson River Virtual Tour,* New York Department of Environmental Conservation, n.d., www.dec.ny.gov/lands/66596.html, accessed March 27, 2014.

On the Department of Environmental Conservation's blueprint for helping the Hudson River Valley ecosystem, see *Hudson River Estuary Area of Biological Concern,* New York Department of Environmental Conservation, n.d., www.dec.ny.gov/docs/remediation_hudson_pdf/hrebcfII2.pdf, accessed March 27, 2014.

Peter Del Tredici is quoted in Courtney Humphries, "This Is Not a Weed," *Boston Globe,* May 23, 2010.

CHAPTER 8

The following websites contain current and historical material relative to the subject of the chapter: World Wildlife Fund, http://gifts.worldwildlife.org; International Union for the Conservation of Nature, www.iucn.org; Conservation International, www.conservation.org; Nature Conservancy, www.nature.org; Wildlife Conservation Society, www.wcs.org; Center for Biodiversity and Conservation at the American Museum of Natural History, www.amnh.org/our-research/center-for-biodiversity-conservation; United Nations, www.un.org; Ford Foundation, www.fordfoundation.org; Duke University Primate Center website, http://lemur.duke.edu/conservation/madagascar-projects; Project Tiger, http://projecttiger.nic.in.

The essay by the Catholic Family and Human Rights Institute of New York, titled "Ford Foundation: Founder of Modern Population Control," was written by Martin Morse Wooster and published as a "white paper." It can be found at http://c-fam.org/en/white-papers/6584-ford-foundation-founder-of-modern-population-control, and it is available in downloadable .PDF format.

One of us (NE) had the great pleasure of being with Dr. Ian Tattersall on an American Museum Discovery ecotour to Madagascar, where the group viewed lemurs at, among other places, Ranomafana in the late 1980s. NE also went twice to Delhi and Corbett National Park near Ramnagar, India, in the company of Dr. Kumar Krishna. The group's local guide and skilled naturalist on both occasions was Sanjay Chhimwal.

SUGGESTIONS FOR FURTHER READING

We found the following books especially thought-provoking while we were developing our argument on cities and the environment over the past few years: Jeb Brugmann, *Welcome to the Urban Revolution: How Cities Are Changing the World* (New York: HarperCollins, 2009); Joel E. Cohen, *How Many People Can the Earth Support?* (New York: Norton, 1995); Paul Kennedy, *Preparing for the Twenty-First Century* (New York: Random House, 1993); David Owen, *Green Metropolis: Why Living Smaller, Living Closer, and Driving Less Are the Keys to Sustainability* (New York: Riverhead Books, 2009).

ILLUSTRATIONS

ACKNOWLEDGMENTS

We are deeply grateful to the library of the American Museum of Natural History: its staff and magnificent holdings have been invaluable resources for us over the years. We also acknowledge our work together on the Hall of Biodiversity at the American Museum—that work was in many ways the precursor to this book.

We acknowledge, too, the helpful insights of the reviewers of our manuscript.

Sid Horenstein: Many of the views and ideas, as well as the explorations of the city, found in this book were in part generated when I prepared numerous short courses for the public on the subjects of geology and nature, and lectures and field trips for the AMNH Education Department and programs for the AMNH Membership Department. Public trips for the New York City Department of Parks and the Fort Tryon Trust, as well as for members of the Bronx County Historical Society, were equally rewarding. For many years I served as a consultant to Time-Life books in the production of geology and natural history titles. My journeys throughout the city and elsewhere were made all the more pleasurable by the accompaniment of my wife, Marcia.

Niles Eldredge: My love and heartfelt thanks to my wife, Michelle, whose diligent editorial work on the manuscript significantly contributed to its completion.

INDEX

Text:	10.75/15 Janson MT Pro
Display:	Janson
Indexer:	Andrew L. Christenson
Compositor:	IDS Infotech, Ltd.
Printer and binder:	Maple Press